W9-CUA-665

Mastering Essential Math Skills

Fractions

Richard W. Fisher

IT IS ILLEGAL TO PHOTOCOPY THIS BOOK

Fractions - Item #405
ISBN 10: 0-9666211-5-8 • ISBN 13: 978-0-9666211-5-0

Copyright © 2008, Richard W. Fisher
1st printing 2008; 2nd printing 2010; 3rd printing 2012; 4th printing 2014

What sets this book apart from other books is its approach. It is not just a math book, but a system of teaching math. Each daily lesson contains four key parts: **Review Exercises**, **Helpful Hints**, **New Material**, and **Problem Solving**. Teachers have flexibility in introducing new topics, but the book provides them with the necessary structure and guidance. The teacher can rest assured that essential math skills in this book are being systematically learned.

This easy-to-follow program requires only fifteen or twenty minutes of instruction per day. Each lesson is concise and self-contained. The daily exercises help students to not only master math skills, but also maintain and reinforce those skills through consistent review - something that is missing in most math programs. Skills learned in this book apply to all areas of the curriculum, and consistent review is built into each daily lesson. Teachers and parents will also be pleased to note that the lessons are quite easy to correct.

This book is based on a system of teaching that was developed by a math instructor over a thirty-year period. This system has produced dramatic results for students. The program quickly motivates students and creates confidence and excitement that leads naturally to success.

Please read the following "How to Use This Book" section and let this program help you to produce dramatic results with your math students.

How to Use This Book

This book is best used on a daily basis. The first lesson should be carefully gone over with students to introduce them to the program and familiarize them with the format. It is hoped that the program will help your students to develop an enthusiasm and passion for math that will stay with them throughout their education.

As you go through these lessons every day, you will soon begin to see growth in the student's confidence, enthusiasm, and skill level. The students will maintain their mastery through the daily review.

Step 1

The students are to complete the review exercises, showing all their work. After completing the problems, it is important for the teacher or parent to go over this section with the students to ensure understanding.

Step 2

Next comes the new material. Use the "Helpful Hints" section to help introduce the new material. Be sure to point out that it is often helpful to come back to this section as the students work independently. This section often has examples that are very helpful to the students.

Step 3

It is highly important for the teacher to work through the two sample problems with the students before they begin to work independently. Working these problems together will ensure that the students understand the topic, and prevent a lot of unnecessary frustration. The two sample problems will get the students off to a good start and will instill confidence as the students begin to work independently.

Step 4

Each lesson has problem solving as the last section of the page. It is recommended that the teacher go through this section, discussing key words and phrases, and also possible strategies. Problem solving is neglected in many math programs, and just a little work each day can produce dramatic results.

Step 5

Solutions are located in the back of the book. Teachers may correct the exercises if they wish, or have the students correct the work themselves.

Table of Contents

Review Exercises

Note to the students and teachers: This section will include problems from all topics covered in this book. Here are some simple problems to get started.

1. $35 + 21 + 16 =$ **2.** $317 + 23 + 209 =$ **3.** $\begin{array}{r} 63 \\ 729 \\ + 472 \\ \hline \end{array}$

4. $715 - 387 =$ **5.** $\begin{array}{r} 621 \\ - 173 \\ \hline \end{array}$ **6.** $\begin{array}{r} 400 \\ - 76 \\ \hline \end{array}$

| **Helpful Hints** | A fraction is a number that names a part of a whole or group. | 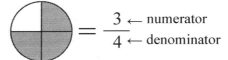 $= \dfrac{3}{4} \begin{array}{l} \leftarrow \text{numerator} \\ \leftarrow \text{denominator} \end{array}$ Think of $\dfrac{3}{4}$ as $\dfrac{3 \text{ of}}{4 \text{ equal parts}}$ |

Write a fraction for each shaded part.
Then write a fraction for each unshaded (white) part.

S1.  S2. 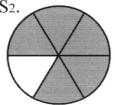 1. 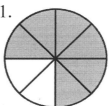 2. 	1. ___
	2. ___
3. 4. 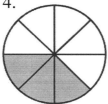 5. 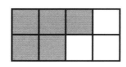 6.	3. ___
	4. ___
7.  8. ⊗ 9. ▦ 10. 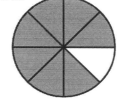	5. ___
	6. ___
	7. ___
	8. ___
	9. ___
	10. ___
	Score

Problem Solving	Crayons come in boxes of 24. How many crayons are there in fifteen boxes?

Review Exercises

1. 24 + 72 + 31 =

2. 705 - 76 =

3. 712
   - 176

4. 55
   666
   + 777

5. Find the difference
   of 752 and 317.

6. Find the sum of
   27, 29, 53, and 64.

Helpful Hints

Use what you have learned to solve the following problems.
* Some fractions may have more than one name.

Example:

This shaded part can be written as $\frac{1}{2}$ and $\frac{2}{4}$.

Write the fraction for each shaded part.
Then write a fraction for each unshaded (white) part.

			1. _____
S1.	S2. 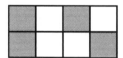	1. 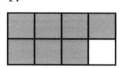	2. _____
			3. _____
			4. _____

2.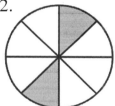

3.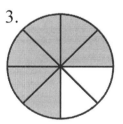

4.

5.

6.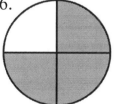

5. _____

6. _____

7. _____

8. _____

7.

8.

9.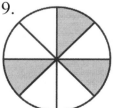

10.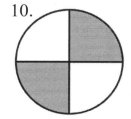

9. _____

10. _____

Score

Problem Solving

Roger earned $850. If he spent $79 for groceries, how much of his earnings were left?

Review Exercises

Write a fraction for each shaded part. Some may have more than one name.

1.

2.

3.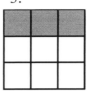

4.

5.

6.

Helpful Hints	There are many ways to sketch a given fraction.	**Examples:** $\dfrac{1}{3} =$ 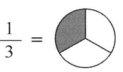 or or

Make a sketch for each fraction. Write the fraction in words in the answer column.

Example: $\dfrac{2}{5}$ = two fifths

S1. $\dfrac{1}{5}$ S2. $\dfrac{3}{8}$ 1. $\dfrac{3}{4}$ 2. $\dfrac{2}{9}$

3. $\dfrac{4}{5}$ 4. $\dfrac{4}{7}$ 5. $\dfrac{7}{12}$ 6. $\dfrac{7}{8}$

7. $\dfrac{4}{8}$ 8. $\dfrac{2}{3}$ 9. $\dfrac{5}{6}$ 10. $\dfrac{4}{6}$

1.
2.
3.
4.
5.
6.
7.
8.
9.
10.

Score

Problem Solving

A car traveled 55 miles per hour for eight hours. How far did the car travel?

Review Exercises

1.
```
  127
   14
+ 316
```

2. Sketch a figure for $\frac{5}{6}$

3. Write a fraction for the shaded part.

4. 501 - 36 =

5.
```
  953
- 271
```

6. 37 + 128 + 27 =

Helpful Hints	Use what you have learned to solve the following problems. * Some fractions may have more than one name.	**Examples:** $\frac{3}{4}$ = 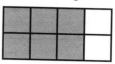 or or

Make a sketch for each fraction. Then write the fraction in words in the answer column.

Example: $\frac{3}{8}$ = three eights

S1. $\frac{1}{3}$ S2. $\frac{3}{10}$ 1. $\frac{5}{8}$ 2. $\frac{1}{6}$

3. $\frac{1}{8}$ 4. $\frac{3}{12}$ 5. $\frac{2}{3}$ 6. $\frac{7}{9}$

7. $\frac{2}{7}$ 8. $\frac{6}{9}$ 9. $\frac{9}{10}$ 10. $\frac{2}{4}$

1.
2.
3.
4.
5.
6.
7.
8.
9.
10.

Score

Problem Solving	Maya is 16 years older than her sister. If Maya's sister is 11, how old is Maya?

Review Exercises

1. Write two fractions for the shaded part.

2. Write two fractions for the shaded part.

3. Sketch a figure for $\frac{2}{3}$.

4. $552 + 78 + 162 =$

5. $700 - 263 =$

6. Find the difference between 752 and 385.

Helpful Hints

$\frac{2}{4}$ has been reduced to its simplest form which is $\frac{1}{2}$. To do so, divide the numerator and denominator by the largest possible number. $= \frac{2}{4} = \frac{1}{2}$

Examples: $2 \overline{\smash{\big)}\frac{6}{8}} = \frac{3}{4}$

Sometimes more than one step can be used:

$2 \overline{\smash{\big)}\frac{24}{28}} = 2 \overline{\smash{\big)}\frac{12}{14}} = \frac{6}{7}$

Reduce each fraction to its lowest terms.

S1. $\frac{5}{10}$ S2. $\frac{12}{16}$ 1. $\frac{12}{15}$ 2. $\frac{15}{20}$

3. $\frac{10}{20}$ 4. $\frac{20}{25}$ 5. $\frac{12}{18}$ 6. $\frac{16}{24}$

7. $\frac{24}{40}$ 8. $\frac{20}{32}$ 9. $\frac{15}{18}$ 10. $\frac{18}{24}$

1.
2.
3.
4.
5.
6.
7.
8.
9.
10.
Score

Problem Solving

If a car can go 32 miles per gallon of gas, how many gallons will the car use traveling 384 miles?

Review Exercises

1. Reduce $\frac{6}{8}$ to its simplest form.

2. Reduce $\frac{20}{25}$ to its simplest form.

3. $755 - 666 =$

4. Sketch a figure for $\frac{7}{8}$.

5. $\begin{array}{r} 764 \\ 44 \\ +\ 555 \\ \hline \end{array}$

6. Find the sum of 6,123 and 7,697.

Helpful Hints	Sometimes fractions can be reduced to simplest form in one step. However, it is okay to use multiple steps.	Examples: $12\overline{)\frac{24}{36}} = \frac{2}{3}$ $2\overline{)\frac{24}{36}} = 2\overline{)\frac{12}{18}} = 2\overline{)\frac{6}{9}} = \frac{2}{3}$

Reduce each fraction to its lowest terms.

S1. $\frac{15}{20}$ S2. $\frac{20}{24}$ 1. $\frac{12}{20}$ 2. $\frac{30}{40}$

3. $\frac{16}{30}$ 4. $\frac{14}{16}$ 5. $\frac{75}{100}$ 6. $\frac{30}{48}$

7. $\frac{30}{32}$ 8. $\frac{25}{30}$ 9. $\frac{80}{90}$ 10. $\frac{40}{48}$

1.

2.

3.

4.

5.

6.

7.

8.

9.

10.

Problem Solving

Juan planned a 3-day, 50 mile hike. If he hikes 17 miles the first day and 19 the second day, how far must he hike on the third day?

Score

Review Exercises

1.   715
    - 367

2. $335 + 36 + 418 =$

3. $7000 - 763 =$

4. Reduce $\dfrac{12}{16}$ to its simplest form.

5.   127
       14
    + 316

6. Reduce $\dfrac{20}{24}$ to its simplest form.

Helpful Hints

Equivalent fractions have the same value. Sometimes it is necessary to write a fraction as an equivalent fraction. Also, any fraction with the same numerator and denominator equals 1. We can multiply any fraction by 1 and not change its value.

Examples: $\dfrac{2}{2}$ or $\dfrac{5}{5}$ or $\dfrac{10}{10}$ all equal 1.

Examples: $\dfrac{2}{3} \times \dfrac{5}{5} = \dfrac{10}{15}$ So, $\dfrac{2}{3} = \dfrac{10}{15}$.

It is easy to solve the problem below. To find the missing numerator, multiply $\dfrac{2}{6}$ by $\dfrac{3}{3}$.

$\dfrac{2}{6} = \dfrac{}{18}$ $\dfrac{2}{6} \times \dfrac{3}{3} = \dfrac{6}{18}$ So, $\dfrac{2}{6} = \dfrac{6}{18}$.   The missing numerator is 6.

Find the missing numerators.

S1. $\dfrac{3}{4} = \dfrac{}{20}$

S2. $\dfrac{3}{4} = \dfrac{}{12}$

1. $\dfrac{2}{7} = \dfrac{}{35}$

2. $\dfrac{3}{4} = \dfrac{}{16}$

3. $\dfrac{2}{3} = \dfrac{}{27}$

4. $\dfrac{11}{12} = \dfrac{}{24}$

5. $\dfrac{7}{10} = \dfrac{}{50}$

6. $\dfrac{9}{14} = \dfrac{}{28}$

7. $\dfrac{3}{13} = \dfrac{}{26}$

8. $\dfrac{2}{3} = \dfrac{}{60}$

9. $\dfrac{4}{9} = \dfrac{}{36}$

10. $\dfrac{11}{15} = \dfrac{}{45}$

1.
2.
3.
4.
5.
6.
7.
8.
9.
10.
Score

Problem Solving

A school spent $2,550.00 on 85 math books.
How much did each book cost?

Review Exercises

1. Write a fraction for the shaded part.

2.
$$629$$
$$- \ 348$$

3. Write two fractions for the shaded portions.

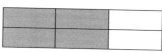

4. $36 + 93 + 42 =$

5. Find the missing numerator.
$$\frac{7}{8} = \frac{}{24}$$

6. $8,000 - 715 =$

Helpful Hints

Use what you have learned to solve the following problems.

Find the missing numerators.

S1. $\dfrac{9}{10} = \dfrac{}{20}$	S2. $\dfrac{15}{18} = \dfrac{}{36}$

1. $\dfrac{4}{15} = \dfrac{}{45}$ 2. $\dfrac{11}{20} = \dfrac{}{80}$

3. $\dfrac{7}{8} = \dfrac{}{40}$ 4. $\dfrac{15}{16} = \dfrac{}{32}$ 5. $\dfrac{3}{7} = \dfrac{}{42}$ 6. $\dfrac{3}{13} = \dfrac{}{39}$

7. $\dfrac{2}{3} = \dfrac{}{9}$ 8. $\dfrac{11}{12} = \dfrac{}{36}$ 9. $\dfrac{4}{7} = \dfrac{}{56}$ 10. $\dfrac{7}{9} = \dfrac{}{63}$

1.

2.

3.

4.

5.

6.

7.

8.

9.

10.

Score

Problem Solving

A car traveled 78 miles on 2 gallons of gas.
How many miles per gallon did the car average?

Review Exercises

1. Write a fraction for the shaded part.

2. Reduce $\dfrac{6}{9}$ to its lowest terms.

3. Reduce $\dfrac{28}{50}$ to its lowest terms.

5.
$$\begin{array}{r} 352 \\ 463 \\ 721 \\ +\ 445 \\ \hline \end{array}$$

4.
$$\begin{array}{r} 610 \\ -\ 375 \\ \hline \end{array}$$

6. Write three fractions for the shaded part.

Helpful Hints

An improper fraction has a numerator that is greater than or equal to its denominator. An improper fraction can be written either as a whole number or as a mixed numeral (a whole number and a fraction).

* Divide the numerator by the denominator.

$$2\overline{\big)7} \begin{array}{l} 3 \\ \underline{6} \\ 1 \end{array} = 3\tfrac{1}{2}$$

Example: 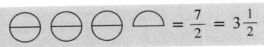 $= \dfrac{7}{2} = 3\dfrac{1}{2}$

Change each improper fraction to a mixed numeral or whole number.
Reduce answers to the lowest terms.

S1. $\dfrac{5}{4} =$

S2. $\dfrac{9}{6} =$

1. $\dfrac{11}{4} =$

2. $\dfrac{11}{7} =$

3. $\dfrac{45}{15} =$

4. $\dfrac{22}{4} =$

5. $\dfrac{37}{12} =$

6. $\dfrac{24}{6} =$

7. $\dfrac{55}{10} =$

8. $\dfrac{40}{6} =$

9. $\dfrac{18}{8} =$

10. $\dfrac{27}{5} =$

1.	
2.	
3.	
4.	
5.	
6.	
7.	
8.	
9.	
10.	
Score	

Problem Solving

On Saturday 12,034 people visited the museum. On Sunday there were 15,768 visitors. How many more people visited the museum on Sunday than on Saturday?

Review Exercises

1. Change $\dfrac{11}{7}$ to a mixed numeral.

2. Change $\dfrac{27}{3}$ to a whole number.

3. Reduce $\dfrac{25}{30}$ to its lowest terms.

4. Give two names for the shaded part.

5. Find the sum of 2,563 and 3,699.

6. Find the difference between 9,112 and 7,132.

Helpful Hints

Use what you have learned to solve the following problems.
For each mixed number, be sure that the fraction is reduced to its lowest terms.

Change each improper fraction to a mixed numeral or whole number.
Reduce answers to the lowest terms.

S1. $\dfrac{19}{15} =$ S2. $\dfrac{65}{20} =$ 1. $\dfrac{20}{8} =$ 2. $\dfrac{9}{8} =$	1.
	2.
	3.
3. $\dfrac{27}{5} =$ 4. $\dfrac{32}{6} =$ 5. $\dfrac{25}{4} =$ 6. $\dfrac{30}{12} =$	4.
	5.
	6.
	7.
7. $\dfrac{85}{10} =$ 8. $\dfrac{16}{9} =$ 9. $\dfrac{96}{15} =$ 10. $\dfrac{40}{24} =$	8.
	9.
	10.
	Score

Problem Solving

A family of 2 adults and 4 children went to the movies. If adult's tickets are $9 each and children's tickets are $6 each, how much was the total cost of the tickets?

Review Exercises

1. Change $\frac{9}{5}$ to a mixed numeral.

2. Reduce $\frac{16}{20}$ to its lowest terms.

3. Change $\frac{40}{18}$ to a mixed numeral.

4. Find the missing numerator.

$$\frac{7}{8} = \frac{}{24}$$

5. Write two fractions for the shaded part.

6. Find the missing numerator.

$$\frac{}{12} = \frac{2}{3}$$

Helpful Hints	To change a mixed number to an improper fraction, do the following: 1. Multiply the denominator and the whole number. 2. Add the answer to the numerator. **Example:** $3\frac{2}{5}$ 2. Add 1. Multiply $= \frac{(15+2)}{5} = \frac{11}{7}$

Change each mixed numeral to an improper fraction.

S1. $2\frac{1}{2} =$

S2. $6\frac{1}{3} =$

1. $7\frac{1}{5} =$

2. $3\frac{2}{3} =$

3. $5\frac{1}{4} =$

4. $4\frac{3}{4} =$

5. $2\frac{5}{6} =$

6. $7\frac{1}{8} =$

7. $7\frac{2}{3} =$

8. $8\frac{3}{4} =$

9. $9\frac{1}{4} =$

10. $1\frac{7}{8} =$

1.
2.
3.
4.
5.
6.
7.
8.
9.
10.
Score

Problem Solving

Lucy plans to drive 360 miles. If her car can travel 20 miles per gallon of gas, how many gallons of gas will she need to make the trip?

Review Exercises

1. Find the missing numerator.

$$\frac{11}{20} = \frac{}{60}$$

2. Find the missing numerator.

$$\frac{}{12} = \frac{3}{4}$$

3. Change $\frac{15}{7}$ to a mixed numeral.

4. Reduce $\frac{20}{25}$ to its lowest terms.

5. Draw a sketch for $\frac{5}{6}$.

6. Reduce $\frac{20}{30}$ to its lowest terms.

Helpful Hints　　Use what you have learned to solve the following problems.

Change each mixed numeral to an improper fraction.

S1. $8\frac{1}{2} =$　　　S2. $4\frac{3}{4} =$　　　1. $7\frac{1}{2} =$　　　2. $5\frac{2}{3} =$

3. $5\frac{3}{8} =$　　　4. $2\frac{2}{3} =$　　　5. $9\frac{1}{3} =$　　　6. $7\frac{3}{5} =$

7. $7\frac{2}{3} =$　　　8. $16\frac{1}{2} =$　　　9. $4\frac{3}{5} =$　　　10. $5\frac{3}{4} =$

1.	
2.	
3.	
4.	
5.	
6.	
7.	
8.	
9.	
10.	
Score	

Problem Solving　　Maria needs 32 cupcakes for a party. If cupcakes come in packages of 6, how many packages must Maria buy?

Review Exercises

1. Change $\dfrac{10}{7}$ to a mixed numeral.

2. Change $4\dfrac{3}{5}$ to an improper fraction.

3. Reduce $\dfrac{25}{35}$ to its lowest terms.

4. Change $5\dfrac{1}{4}$ to an improper fraction.

5. Change $\dfrac{20}{15}$ to a mixed numeral.

6. Draw a sketch for $\dfrac{2}{3}$.

Helpful Hints

To add fractions with like denominators, add the numerators and then ask the following questions:

1. Is the fraction improper? If it is, make it a mixed numeral or whole number.

2. Can the fraction be reduced? If it can, reduce it to its simplest form.

Example:

$$\dfrac{7}{10} + \dfrac{5}{10} = \dfrac{12}{10} = 1\dfrac{2}{10} = 1\dfrac{1}{5}$$

S1. $\dfrac{1}{10} + \dfrac{7}{10}$

S2. $\dfrac{7}{8} + \dfrac{3}{8}$

1. $\dfrac{5}{9} + \dfrac{2}{9}$

2. $\dfrac{1}{8} + \dfrac{5}{8}$

3. $\dfrac{7}{8} + \dfrac{5}{8}$

4. $\dfrac{7}{10} + \dfrac{2}{10}$

5. $\dfrac{5}{8} + \dfrac{5}{8}$

6. $\dfrac{9}{16} + \dfrac{11}{16}$

7. $\dfrac{2}{5} + \dfrac{4}{5} + \dfrac{3}{5}$

8. $\dfrac{3}{8} + \dfrac{5}{8} + \dfrac{2}{8}$

9. $\dfrac{9}{10} + \dfrac{7}{10}$

10. $\dfrac{5}{6} + \dfrac{5}{6}$

1.
2.
3.
4.
5.
6.
7.
8.
9.
10.
Score

Problem Solving

A recipe calls for $\dfrac{3}{8}$ cup of flour and $\dfrac{7}{8}$ cup of sugar. How much flour and sugar is needed altogether?

Review Exercises

1. Change $5\frac{1}{2}$ to an improper fraction.

2. Change $\frac{16}{5}$ to a mixed numeral.

3. $\frac{3}{5}$

   $+\ \frac{1}{5}$

4. $\frac{3}{4}$

   $+\ \frac{3}{4}$

5. Make a sketch for $\frac{7}{8}$.

6. Reduce $\frac{18}{20}$ to its lowest terms.

Helpful Hints	Use what you have learned to solve the following problems. * Remember: 1. Is the fraction improper? If it is, make it a mixed numeral or whole number. 2. Can the fraction be reduced? If it can, reduce it to its simplest form.

Change each mixed number to an improper fraction.

S1. $\frac{7}{10}$ S2. $\frac{11}{12}$ 1. $\frac{1}{8}$ 2. $\frac{3}{10}$

    $+\ \frac{1}{10}$ $+\ \frac{7}{12}$ $+\ \frac{5}{8}$ $+\ \frac{4}{10}$

3. $\frac{9}{10}$ 4. $\frac{11}{16}$ 5. $\frac{7}{12}$ 6. $\frac{3}{4}$

    $+\ \frac{5}{10}$ $+\ \frac{7}{16}$ $+\ \frac{6}{12}$ $\frac{1}{4}$

                                                                         $+\ \frac{3}{4}$

7. $\frac{5}{6}$ 8. $\frac{9}{13}$ 9. $\frac{2}{3}$ 10. $\frac{11}{15}$

    $+\ \frac{5}{6}$ $+\ \frac{7}{13}$ $\frac{2}{3}$ $+\ \frac{7}{15}$

                                             $+\ \frac{2}{3}$

1.
2.
3.
4.
5.
6.
7.
8.
9.
10.

Score

Problem Solving	Saturday it rained $\frac{7}{10}$ inches and Sunday it rained $\frac{9}{10}$ inches. What is the total amount of rain that fell during the two days?

Review Exercises

1. $\dfrac{5}{8}$
$+\dfrac{4}{8}$

2. $\dfrac{3}{10}$
$+\dfrac{3}{10}$

3. Reduce $\dfrac{16}{18}$ to its lowest terms.

6. $\dfrac{3}{4}$
$+\dfrac{1}{4}$

4. Change $\dfrac{21}{4}$ to a mixed numeral.

5. Change $3\dfrac{3}{4}$ to an improper fraction.

| **Helpful Hints** | 1. Add the fractions first.
 2. Add the whole numbers next.
 3. If there is an improper fraction, change it to a mixed numeral.
 4. Add the mixed numeral to the whole number. | **Example:** $6\dfrac{7}{8}$ $+\;2\dfrac{5}{8}$ $8\dfrac{12}{8}=8+1\dfrac{4}{8}=9\dfrac{4}{8}=9\dfrac{1}{2}$ * Reduce the answer to its lowest terms. |

S1. $4\dfrac{1}{10}$
$+2\dfrac{7}{10}$

S2. $3\dfrac{7}{8}$
$+4\dfrac{3}{8}$

1. $5\dfrac{4}{7}$
$+2\dfrac{2}{7}$

2. $3\dfrac{7}{10}$
$+2\dfrac{5}{10}$

3. $5\dfrac{3}{4}$
$+6\dfrac{1}{4}$

4. $3\dfrac{5}{6}$
$+2\dfrac{3}{6}$

5. $4\dfrac{5}{7}$
$+2\dfrac{4}{7}$

6. $2\dfrac{3}{4}$
$+4\dfrac{3}{4}$

7. $5\dfrac{9}{10}$
$+4\dfrac{3}{10}$

8. $2\dfrac{11}{15}$
$+3\dfrac{6}{15}$

9. $6\dfrac{3}{10}$
$+3\dfrac{7}{10}$

10. $9\dfrac{3}{5}$
$+2\dfrac{3}{5}$

| 1. |
| 2. |
| 3. |
| 4. |
| 5. |
| 6. |
| 7. |
| 8. |
| 9. |
| 10. |
| Score |

Problem Solving

A baker uses $\dfrac{3}{4}$ cups of flour for each pie he bakes. He uses $\dfrac{1}{4}$ cups of flour for each cake. How much flour is used to make two pies and one cake?

Review Exercises

1. $\frac{1}{3}$
   $+ \frac{1}{3}$
   _____

2. $\frac{7}{10}$
   $+ \frac{1}{10}$
   _____

3. $\frac{14}{15}$
   $+ \frac{4}{15}$
   _____

4. Reduce $\frac{30}{32}$ to its lowest terms.

5. Change $3\frac{5}{6}$ to an improper fraction.

6. Change $\frac{17}{3}$ to a mixed numeral.

Helpful Hints

Use what you have learned to solve the following problems.
* Remember:
1. If there is an improper fraction, change it to a mixed numeral.
2. Reduce each fraction to its lowest terms.

Change each mixed number to an improper fraction.

S1. $3\frac{7}{10}$
    $+ 4\frac{1}{10}$
    _____

S2. $5\frac{7}{10}$
    $+ 3\frac{9}{10}$
    _____

1. $3\frac{1}{5}$
   $+ 2\frac{3}{5}$
   _____

2. $4\frac{4}{5}$
   $+ 3\frac{3}{5}$
   _____

3. $9\frac{1}{6}$
   $+ 5\frac{5}{6}$
   _____

4. $7\frac{7}{8}$
   $+ 5\frac{5}{8}$
   _____

5. $13\frac{3}{5}$
   $+ 16\frac{3}{5}$
   _____

6. $9\frac{2}{3}$
   $+ 7\frac{2}{3}$
   _____

7. $3\frac{7}{15}$
   $+ 4\frac{9}{15}$
   _____

8. $2\frac{3}{16}$
   $+ 3\frac{5}{16}$
   _____

9. $7\frac{7}{10}$
   $+ 3\frac{3}{10}$
   _____

10. $7\frac{7}{11}$
    $+ 5\frac{9}{11}$
    _____

1.
2.
3.
4.
5.
6.
7.
8.
9.
10.

Score

Problem Solving

Anna rides her bike $\frac{7}{8}$ miles to school. If she has already ridden $\frac{3}{8}$ miles, how much farther must she ride before she gets to school?

Review Exercises

1. Change $\dfrac{18}{12}$ to a mixed numeral.

2. Reduce $\dfrac{20}{25}$ to its lowest terms.

3. $\begin{array}{r} \dfrac{3}{5} \\ + \dfrac{4}{5} \\ \hline \end{array}$

4. $\begin{array}{r} \dfrac{7}{8} \\ + \dfrac{3}{8} \\ \hline \end{array}$

5. $\begin{array}{r} \dfrac{3}{10} \\ + \dfrac{3}{10} \\ \hline \end{array}$

6. $\begin{array}{r} 5\dfrac{6}{7} \\ + 3\dfrac{5}{7} \\ \hline \end{array}$

Helpful Hints

Use what you have learned to solve the following problems.
* Remember:
1. Change improper fractions to a mixed numerals.
2. Reduce each fraction to its lowest terms.

S1. $\begin{array}{r} \dfrac{7}{8} \\ + \dfrac{3}{8} \\ \hline \end{array}$

S2. $\begin{array}{r} 4\dfrac{4}{5} \\ + 3\dfrac{2}{5} \\ \hline \end{array}$

1. $\begin{array}{r} \dfrac{3}{6} \\ + \dfrac{1}{6} \\ \hline \end{array}$

2. $\begin{array}{r} 4\dfrac{5}{8} \\ + 3\dfrac{3}{8} \\ \hline \end{array}$

3. $\begin{array}{r} \dfrac{9}{10} \\ + \dfrac{7}{10} \\ \hline \end{array}$

4. $\begin{array}{r} 5\dfrac{2}{3} \\ + 3\dfrac{2}{3} \\ \hline \end{array}$

5. $\begin{array}{r} 7\dfrac{7}{8} \\ + 3\dfrac{5}{8} \\ \hline \end{array}$

6. $\begin{array}{r} \dfrac{2}{3} \\ \dfrac{1}{3} \\ + \dfrac{2}{3} \\ \hline \end{array}$

7. $\begin{array}{r} \dfrac{5}{16} \\ + \dfrac{5}{16} \\ \hline \end{array}$

8. $\begin{array}{r} 9\dfrac{7}{9} \\ + 3\dfrac{2}{9} \\ \hline \end{array}$

9. $\begin{array}{r} \dfrac{5}{8} \\ + \dfrac{5}{8} \\ \hline \end{array}$

10. $\begin{array}{r} 2\dfrac{11}{12} \\ + 3\dfrac{5}{12} \\ \hline \end{array}$

1.
2.
3.
4.
5.
6.
7.
8.
9.
10.
Score

Problem Solving

Pablo worked $7\dfrac{3}{4}$ hours on Monday and $6\dfrac{3}{4}$ hours on Tuesday. How many hours did he work altogether?

Review Exercises

1. Draw a sketch for $\frac{5}{6}$.

2. Change $\frac{7}{2}$ to a mixed numeral.

3. Reduce $\frac{14}{20}$ to its lowest terms.

4. $\frac{5}{6}$
$+ \frac{3}{6}$

5. $\frac{9}{10}$
$+ \frac{1}{10}$

6. $7\frac{3}{4}$
$+ 8\frac{3}{4}$

Helpful Hints

Use what you have learned to solve the following problems.

S1. $4\frac{5}{7}$
$+ 3\frac{1}{7}$

S2. $\frac{11}{12}$
$+ \frac{7}{12}$

1. $\frac{3}{8}$
$+ \frac{1}{8}$

2. $5\frac{5}{8}$
$+ 3\frac{5}{8}$

3. $\frac{7}{8}$
$\frac{3}{8}$
$+ \frac{5}{8}$

4. $9\frac{2}{3}$
$+ 3\frac{2}{3}$

5. $\frac{7}{10}$
$+ \frac{1}{10}$

6. $6\frac{7}{10}$
$+ 3\frac{5}{10}$

7. $\frac{11}{12}$
$+ \frac{3}{12}$

8. $4\frac{2}{3}$
$3\frac{1}{3}$
$+ 5\frac{2}{3}$

9. $3\frac{3}{8}$
$4\frac{7}{8}$
$+ 3\frac{5}{8}$

10. $\frac{7}{9}$
$+ \frac{3}{9}$

1.

2.

3.

4.

5.

6.

7.

8.

9.

10.

Score

Problem Solving

Pierre scores of 86, 88, and 96 on his tests. What was his average score?

Review Exercises

1. $\frac{3}{8}$
    $+ \frac{1}{8}$

2. $\frac{5}{7}$
    $+ \frac{4}{7}$

3. $\frac{9}{16}$
    $+ \frac{11}{16}$

4. $3\frac{1}{4}$
    $+ 4\frac{3}{4}$

5. $8\frac{3}{10}$
    $+ 7\frac{5}{10}$

6. $4\frac{3}{4}$
    $+ 5\frac{3}{4}$

Helpful Hints	To subtract fractions that have like denominators, subtract the numerators. Reduce the fractions to their lowest terms.	**Example:** $\frac{9}{10} - \frac{3}{10} = \frac{6}{10} = \frac{3}{5}$

S1. $\frac{3}{4} - \frac{1}{4}$

S2. $\frac{11}{16} - \frac{1}{16}$

1. $\frac{7}{10} - \frac{1}{10}$

2. $\frac{4}{5} - \frac{1}{5}$

3. $\frac{7}{8} - \frac{1}{8}$

4. $\frac{17}{20} - \frac{2}{20}$

5. $\frac{17}{18} - \frac{3}{18}$

6. $\frac{11}{12} - \frac{5}{12}$

7. $\frac{24}{25} - \frac{4}{25}$

8. $\frac{11}{35} - \frac{1}{35}$

9. $\frac{11}{12} - \frac{2}{12} =$

10. $\frac{3}{10} - \frac{1}{10}$

1.
2.
3.
4.
5.
6.
7.
8.
9.
10.
Score

Problem Solving Mrs. Lopez had $\frac{7}{8}$ pounds of sugar. If she used $\frac{3}{8}$ pounds in baking a cake, how many pounds of sugar were left?

Review Exercises

1. Reduce $\frac{20}{25}$ to its lowest terms.

2. $3\frac{3}{5}$
   $+ 5\frac{4}{5}$

3. Change $7\frac{2}{3}$ to an improper fraction.

4. $\frac{9}{10}$
   $- \frac{1}{10}$

5. $\frac{4}{5} + \frac{3}{5} =$

6. Change $\frac{19}{7}$ to a mixed numeral.

Helpful Hints	Use what you have learned to solve the following problems. Remember to reduce answers to lowest terms.

S1. $\frac{7}{12}$
   $- \frac{1}{12}$

S2. $\frac{29}{32}$
   $- \frac{1}{32}$

1. $\frac{15}{16}$
   $- \frac{3}{16}$

2. $\frac{14}{15}$
   $- \frac{2}{15}$

3. $\frac{7}{8} - \frac{5}{8} =$

4. $\frac{11}{24}$
   $- \frac{3}{24}$

5. $\frac{19}{20}$
   $- \frac{4}{20}$

6. $\frac{11}{18}$
   $- \frac{5}{18}$

7. $\frac{2}{3}$
   $- \frac{1}{3}$

8. $\frac{23}{35}$
   $- \frac{2}{35}$

9. $\frac{49}{50}$
   $- \frac{9}{50}$

10. $\frac{13}{48}$
   $- \frac{1}{48}$

1.
2.
3.
4.
5.
6.
7.
8.
9.
10.

Score

Problem Solving	Fredo worked $3\frac{1}{4}$ hours on Monday, $2\frac{3}{4}$ hours on Tuesday, and $4\frac{3}{4}$ hours on Wednesday. How many hours did he work altogether?

Review Exercises

1. $\begin{array}{r} \frac{3}{4} \\ -\ \frac{1}{4} \\ \hline \end{array}$

2. $\begin{array}{r} 2\frac{2}{3} \\ +\ 7\frac{1}{3} \\ \hline \end{array}$

3. $\begin{array}{r} \frac{7}{12} \\ -\ \frac{4}{12} \\ \hline \end{array}$

4. $\begin{array}{r} 4\frac{5}{8} \\ +\ 3\frac{3}{8} \\ \hline \end{array}$

5. $\begin{array}{r} 12\frac{2}{3} \\ +\ 13\frac{2}{3} \\ \hline \end{array}$

6. $\begin{array}{r} \frac{15}{16} \\ +\ \frac{11}{16} \\ \hline \end{array}$

Helpful Hints

To subtract mixed numerals with like denominators, subtract the fractions first, then the whole numbers. Reduce fractions to lowest terms. If the fraction can't be subtracted, take one from the whole number, increase the fraction, then subtract.

Examples:

$\begin{array}{r} 7\frac{3}{4} \\ -\ 2\frac{1}{4} \\ \hline 6\frac{2}{4} = 6\frac{1}{2} \end{array}$

$6\ \cancel{7}\frac{1}{4} + \frac{4}{4} = \frac{5}{4}$

$\begin{array}{r} -\ 2\frac{3}{4} \\ \hline 4\frac{2}{4} = 4\frac{1}{2} \end{array}$

S1. $\begin{array}{r} 4\frac{3}{4} \\ -\ 1\frac{1}{4} \\ \hline \end{array}$

S2. $\begin{array}{r} 5\frac{1}{5} \\ -\ 2\frac{3}{5} \\ \hline \end{array}$

1. $\begin{array}{r} 6\frac{3}{8} \\ -\ 2\frac{1}{8} \\ \hline \end{array}$

2. $\begin{array}{r} 7\frac{1}{4} \\ -\ 2\frac{3}{4} \\ \hline \end{array}$

3. $\begin{array}{r} 8\frac{3}{8} \\ -\ 4\frac{7}{8} \\ \hline \end{array}$

4. $\begin{array}{r} 7\frac{7}{8} \\ -\ 2\frac{1}{8} \\ \hline \end{array}$

5. $\begin{array}{r} 4\frac{3}{10} \\ -\ 1\frac{7}{10} \\ \hline \end{array}$

6. $\begin{array}{r} 5\frac{3}{20} \\ -\ 2\frac{11}{20} \\ \hline \end{array}$

7. $\begin{array}{r} 5\frac{3}{5} \\ -\ 2\frac{3}{5} \\ \hline \end{array}$

8. $\begin{array}{r} 12\frac{4}{9} \\ -\ 10\frac{7}{9} \\ \hline \end{array}$

9. $\begin{array}{r} 7\frac{1}{7} \\ -\ 2\frac{3}{7} \\ \hline \end{array}$

10. $\begin{array}{r} 8\frac{1}{15} \\ -\ 2\frac{7}{15} \\ \hline \end{array}$

1.	
2.	
3.	
4.	
5.	
6.	
7.	
8.	
9.	
10.	
Score	

Problem Solving

Jose and Hector are taking a $20\frac{1}{4}$ mile bike ride. If they have ridden $8\frac{3}{4}$ miles, how much farther do they have to ride?

Review Exercises

1. Change $\frac{19}{15}$ to a mixed numeral.

2. $\begin{array}{r} \frac{15}{16} \\ -\ \frac{3}{16} \\ \hline \end{array}$

3. Reduce $\frac{30}{32}$ to its lowest terms.

4. $\begin{array}{r} \frac{3}{7} \\ +\ \frac{5}{7} \\ \hline \end{array}$

5. Write two fractions for the shaded part.

6. $\begin{array}{r} 4\frac{5}{8} \\ +\ 3\frac{5}{8} \\ \hline \end{array}$

Helpful Hints

Use what you have learned to solve the following problems.
* Remember:
1. If necessary take one whole from the whole number, increase the fraction, then subtract.
2. Reduce the answer to lowest terms.

S1. $\begin{array}{r} 5\frac{1}{4} \\ -\ 3\frac{3}{4} \\ \hline \end{array}$

S2. $\begin{array}{r} 9\frac{1}{8} \\ -\ 3\frac{5}{8} \\ \hline \end{array}$

1. $\begin{array}{r} 7\frac{2}{3} \\ -\ 1\frac{1}{3} \\ \hline \end{array}$

2. $\begin{array}{r} 8\frac{1}{3} \\ -\ 2\frac{2}{3} \\ \hline \end{array}$

3. $\begin{array}{r} 5\frac{7}{8} \\ -\ 1\frac{1}{8} \\ \hline \end{array}$

4. $\begin{array}{r} 6\frac{3}{8} \\ -\ 2\frac{7}{8} \\ \hline \end{array}$

5. $\begin{array}{r} 12\frac{7}{11} \\ -\ 9\frac{10}{11} \\ \hline \end{array}$

6. $\begin{array}{r} 7\frac{5}{12} \\ -\ 2\frac{7}{12} \\ \hline \end{array}$

7. $7\frac{3}{4} - 1\frac{1}{4} =$

8. $7\frac{1}{8} - 2\frac{3}{8} =$

9. $\begin{array}{r} 12\frac{1}{20} \\ -\ 7\frac{7}{20} \\ \hline \end{array}$

10. $\begin{array}{r} 7\frac{1}{9} \\ -\ 3\frac{7}{9} \\ \hline \end{array}$

1.

2.

3.

4.

5.

6.

7.

8.

9.

10.

Score

Problem Solving

The Jones family traveled 80 miles in their car. If the car travels 20 miles per gallon of gas, how many gallons of gas did the car use? If gas is $3 per gallon, how much did the gas used for the trip cost?

Review Exercises

1. $\dfrac{7}{8}$
   $-\dfrac{1}{8}$

2. $3\dfrac{19}{20}$
   $-1\dfrac{4}{20}$

3. $7\dfrac{5}{8}$
   $-3\dfrac{7}{8}$

4. $\dfrac{5}{8}$
   $+\dfrac{7}{8}$

5. $3\dfrac{7}{10}$
   $+3\dfrac{1}{10}$

6. $5\dfrac{7}{15}$
   $+3\dfrac{10}{15}$

Helpful Hints

To subtract a fraction or mixed numeral from a whole number, take one from the whole number and make it a fraction, then subtract.

Examples:

$$3\cancel{4} \longrightarrow \dfrac{4}{4}$$
$$-2 \quad\quad \dfrac{1}{4}$$
$$\overline{\quad\quad 1\dfrac{3}{4}}$$

$$6\cancel{7} \longrightarrow \dfrac{5}{5}$$
$$- \quad\quad \dfrac{3}{5}$$
$$\overline{\quad\quad 6\dfrac{2}{5}}$$

S1. 7
   $-3\dfrac{3}{5}$

S2. 9
   $-\dfrac{5}{8}$

1. 12
   $-7\dfrac{1}{8}$

2. 6
   $-3\dfrac{3}{7}$

3. 12
   $-\dfrac{7}{15}$

4. 7
   $-2\dfrac{9}{10}$

5. 15
   $-2\dfrac{1}{8}$

6. 42
   $-29\dfrac{1}{7}$

7. 12
   $-\dfrac{9}{11}$

8. 13
   $-7\dfrac{1}{5}$

9. 53
   $-29\dfrac{3}{7}$

10. 16
   $-\dfrac{7}{15}$

1.	
2.	
3.	
4.	
5.	
6.	
7.	
8.	
9.	
10.	
Score	

Problem Solving

A tailor had 15 yards of cloth. He used $5\dfrac{3}{4}$ yards to make a suit. How many yards of cloth were left?

Review Exercises

1. Make a sketch for $\frac{5}{6}$.

2. $7\frac{1}{3}$
$-2\frac{2}{3}$

3. Change $\frac{23}{4}$ to a mixed numeral.

4. $\frac{7}{8}$
$+\frac{7}{8}$

5. $\frac{9}{10}$
$-\frac{3}{10}$

6. Change $3\frac{4}{5}$ to an improper fraction.

Helpful Hints

Use what you have learned to solve the following problems.

	1.

S1. 9
$-\frac{7}{8}$

S2. 12
$-2\frac{3}{7}$

1. 14
$-9\frac{1}{4}$

2. 61
$-28\frac{3}{8}$

3. 4
$-\frac{15}{16}$

4. 9
$-\frac{15}{24}$

5. 14
$-3\frac{1}{24}$

6. 22
$-6\frac{9}{15}$

7. 40
$-\frac{13}{24}$

8. 28
$-19\frac{13}{24}$

9. 15
$-2\frac{19}{20}$

10. 41
$-16\frac{12}{15}$

1.

2.

3.

4.

5.

6.

7.

8.

9.

10.

Score

Problem Solving

A car's gas tank contained 20 gallons of gas, On Monday $5\frac{1}{4}$ gallons were used, and on Tuesday $3\frac{1}{4}$ gallons were used. How many gallons of gas were left in he tank?

Review Exercises

1. $7\frac{1}{3}$
$-2\frac{2}{3}$

2. $5\frac{7}{15}$
$+3\frac{10}{15}$

3. Change $\frac{31}{25}$ to a mixed numeral.

4. Reduce $\frac{18}{30}$ to its lowest terms.

5. 7
$-2\frac{3}{4}$

6. $\frac{11}{12}+\frac{3}{12}=$

Helpful Hints

Use what you have learned to solve the following problems. Regroup when necessary. Reduce all answers to their lowest terms. If problems are positioned horizontally, put them in columns before working.

S1. $3\frac{9}{10}$
$+2\frac{3}{10}$

S2. $6\frac{1}{5}$
$-2\frac{3}{5}$

1. $\frac{6}{7}$
$+\frac{3}{7}$

2. $\frac{14}{15}$
$-\frac{4}{15}$

3. $7\frac{4}{9}$
$+3\frac{7}{9}$

4. 9
$-2\frac{1}{4}$

5. $7\frac{3}{8}$
$+2\frac{3}{8}$

6. $9-2\frac{1}{2}=$

7. $4\frac{11}{12}$
$+3\frac{5}{12}$

8. $7\frac{1}{3}$
$-2\frac{2}{3}$

9. $12\frac{3}{10}$
$-4\frac{7}{10}$

10. $\frac{9}{10}$
$\frac{7}{10}$
$+\frac{3}{10}$

1.
2.
3.
4.
5.
6.
7.
8.
9.
10.
Score

Problem Solving

A plane traveled 2,700 miles in six hours. What was its average speed per hour?

Review Exercises

1. $5\frac{1}{4}$
$-2\frac{3}{4}$

2. 7
$-\frac{11}{15}$

3. $9\frac{4}{5}$
$+7\frac{3}{5}$

4. Change $\frac{18}{12}$ to a mixed numeral.

5. Reduce $\frac{20}{32}$ to its lowest terms.

6. Change $5\frac{1}{4}$ to an improper fraction.

Helpful Hints

Use what you have learned to solve the following problems. Sometimes it is helpful to refer to the "Helpful Hints" section on previous pages.

S1. $7\frac{1}{8}$
$-2\frac{3}{8}$

S2. $4\frac{5}{8}$
$+3\frac{7}{8}$

1. $\frac{11}{12}$
$-\frac{3}{12}$

2. $\frac{8}{9}$
$+\frac{5}{9}$

3. $6\frac{3}{7}$
$+4\frac{5}{7}$

4. 7
$-\frac{12}{20}$

5. $23\frac{3}{8}$
$+44\frac{7}{8}$

6. $7-3\frac{3}{4}=$

7. $9\frac{9}{10}$
$+4\frac{3}{10}$

8. $6\frac{1}{12}$
$-1\frac{7}{12}$

9. $3\frac{11}{15}$
$+7\frac{4}{15}$

10. $\frac{7}{8}$
$\frac{5}{8}$
$+\frac{4}{8}$

1.
2.
3.
4.
5.
6.
7.
8.
9.
10.
Score

Problem Solving

There are 7 rows of desks in a classroom. There are 12 desks in each row. If 5 of the desks are taken, how many are empty?

Review Exercises

1. Find the sum of $\frac{7}{8}$ and $\frac{3}{8}$.

2. Find the difference between $3\frac{1}{4}$ and $1\frac{3}{4}$.

3. Reduce $\frac{42}{48}$ to its lowest terms.

4. Change $3\frac{2}{5}$ to an improper fraction.

5. $\frac{5}{6} + \frac{3}{6} + \frac{2}{6} =$

6. $7 - 2\frac{3}{4} =$

Helpful Hints

To add or subtract fractions with unlike denominations you need to find the least common denominator (LCD). The LCD is the smallest number, other than zero, that each denominator will divide into evenly.

Examples: The least common denominator of:

$\frac{1}{5}$ and $\frac{1}{10}$ is 10. $\frac{3}{8}$ and $\frac{1}{6}$ is 24.

Find the least common denominator of each of the following.

S1. $\frac{3}{4}$ and $\frac{1}{5}$

S2. $\frac{5}{6}$ and $\frac{3}{8}$

1. $\frac{5}{6}$ and $\frac{1}{9}$

2. $\frac{3}{4}$ and $\frac{1}{12}$

3. $\frac{11}{21}$ and $\frac{1}{7}$

4. $\frac{1}{6}$, $\frac{1}{4}$ and $\frac{3}{8}$

5. $\frac{1}{9}$, $\frac{5}{6}$ and $\frac{7}{12}$

6. $\frac{3}{8}$ and $\frac{1}{7}$

7. $\frac{7}{12}$ and $\frac{5}{48}$

8. $\frac{1}{15}$, $\frac{9}{20}$ and $\frac{1}{6}$

9. $\frac{11}{12}$, $\frac{3}{8}$ and $\frac{11}{48}$

10. $\frac{3}{8}$, $\frac{3}{16}$ and $\frac{1}{12}$

1.	
2.	
3.	
4.	
5.	
6.	
7.	
8.	
9.	
10.	
Score	

Problem Solving

There are 468 students in a school. If they are divided in 18 equally sized classes, how many are in each class?

Review Exercises

1. $\dfrac{4}{5}$
    $+\dfrac{2}{5}$

2. $9\dfrac{5}{8}$
    $+6\dfrac{5}{8}$

3. $7\dfrac{11}{16}$
    $+2\dfrac{3}{16}$

4. $\dfrac{9}{15}$
    $-\dfrac{3}{15}$

5. 7
    $-\dfrac{15}{24}$

6. $7\dfrac{1}{6}$
    $-1\dfrac{5}{6}$

Helpful Hints

Use what you have learned to solve the following problems.

Find the least common denominator of each of the following.

S1. $\dfrac{7}{8}$ and $\dfrac{11}{12}$

S2. $\dfrac{2}{3}$, $\dfrac{5}{12}$ and $\dfrac{5}{6}$

1. $\dfrac{7}{9}$ and $\dfrac{5}{6}$

2. $\dfrac{3}{4}$ and $\dfrac{1}{7}$

3. $\dfrac{1}{2}$ and $\dfrac{7}{15}$

4. $\dfrac{7}{30}$ and $\dfrac{1}{45}$

5. $\dfrac{3}{4}$ and $\dfrac{7}{18}$

6. $\dfrac{1}{9}$ and $\dfrac{1}{12}$

7. $\dfrac{1}{4}$, $\dfrac{1}{2}$ and $\dfrac{1}{3}$

8. $\dfrac{1}{12}$ and $\dfrac{7}{16}$

9. $\dfrac{1}{9}$ and $\dfrac{11}{72}$

10. $\dfrac{1}{2}$, $\dfrac{1}{5}$ and $\dfrac{1}{6}$

| 1. |
| 2. |
| 3. |
| 4. |
| 5. |
| 6. |
| 7. |
| 8. |
| 9. |
| 10. |
| Score |

Problem Solving

Nina bought 12 gallons of paint to paint her house. If she used $8\dfrac{3}{4}$ gallons how much paint does she have left?

Review Exercises

1. $\dfrac{7}{12}$
    $+\dfrac{9}{12}$

2. $\dfrac{15}{16}$
    $-\dfrac{11}{16}$

3. Find the least common denominator for $\dfrac{5}{6}$ and $\dfrac{7}{15}$.

6. $\dfrac{5}{2}$
    $\dfrac{3}{2}$
    $+\dfrac{7}{2}$

4. Reduce $\dfrac{56}{70}$ to its lowest terms.

5. Change $\dfrac{57}{11}$ to a mixed number.

Helpful Hints	To add fractions with unlike denominators, find the least common denominator. Multiply each fraction by one to make equivalent fractions. Finally, add.	**Examples:**

$$\frac{2}{5} \times \frac{2}{2} = \frac{4}{10} \qquad \frac{5}{6} \times \frac{2}{2} = \frac{10}{12}$$
$$+\frac{1}{2} \times \frac{5}{5} = \frac{5}{10} \qquad +\frac{1}{4} \times \frac{3}{3} = \frac{3}{12}$$
$$\frac{9}{10} \qquad\qquad \frac{13}{12} = 1\frac{1}{12}$$

S1. $\dfrac{1}{3}$
     $+\dfrac{1}{4}$

S2. $\dfrac{3}{5}$
     $+\dfrac{7}{10}$

1. $\dfrac{5}{9}$
    $+\dfrac{1}{3}$

2. $\dfrac{2}{3}$
    $+\dfrac{1}{2}$

3. $\dfrac{1}{4}$
    $+\dfrac{2}{3}$

4. $\dfrac{3}{4}$
    $+\dfrac{2}{3}$

5. $\dfrac{5}{6}$
    $+\dfrac{5}{12}$

6. $\dfrac{1}{2}$
    $+\dfrac{3}{4}$

7. $\dfrac{1}{6}$
    $+\dfrac{3}{4}$

8. $\dfrac{7}{9}$
    $+\dfrac{1}{4}$

9. $\dfrac{7}{11}$
    $+\dfrac{1}{2}$

10. $\dfrac{3}{8}$
     $+\dfrac{1}{6}$

1.
2.
3.
4.
5.
6.
7.
8.
9.
10.
Score

Problem Solving

Frankie worked for $7\frac{1}{4}$ hours on Tuesday and $5\frac{3}{4}$ on Wednesday. How many more hours did he work on Tuesday then on Wednesday?

Review Exercises

1. $\dfrac{9}{10}$
$+\dfrac{3}{10}$

2. $3\dfrac{3}{8}$
$+4\dfrac{7}{8}$

3. 5
$-\dfrac{14}{20}$

4. 6
$-3\dfrac{3}{8}$

5. $7\dfrac{1}{4}$
$-2\dfrac{3}{4}$

6. Find the least common denominator for $\dfrac{5}{8}$, $\dfrac{1}{3}$ and $\dfrac{7}{12}$.

Use what you have learned to solve the following problems.
*Remember, Change all improper fractions to mixed numerals. Reduce all fractions to lowest terms.

S1. $\dfrac{2}{3}$
$+\dfrac{1}{9}$

S2. $\dfrac{3}{4}$
$+\dfrac{5}{6}$

1. $\dfrac{7}{12}$
$+\dfrac{1}{3}$

2. $\dfrac{1}{2}$
$\dfrac{1}{4}$
$+\dfrac{2}{5}$

3. $\dfrac{7}{25}$
$+\dfrac{3}{5}$

4. $\dfrac{11}{12}$
$+\dfrac{1}{2}$

5. $\dfrac{11}{15}$
$+\dfrac{1}{3}$

6. $\dfrac{2}{3}$
$+\dfrac{3}{5}$

7. $\dfrac{2}{3}$
$\dfrac{1}{2}$
$+\dfrac{3}{8}$

8. $\dfrac{3}{11}$
$+\dfrac{5}{33}$

9. $\dfrac{5}{12}$
$+\dfrac{3}{8}$

10. $\dfrac{4}{6}$
$+\dfrac{1}{10}$

1.	
2.	
3.	
4.	
5.	
6.	
7.	
8.	
9.	
10.	
Score	

Kareem needs 70 dollars for a new cell phone. On Monday he earned $17\dfrac{1}{2}$ dollars and on Wednesday he earned $16\dfrac{1}{2}$ dollars. How much more does he need to have enough to buy the cell phone?

Review Exercises

1. $\dfrac{2}{5}$
    $+\dfrac{1}{5}$

2. $\dfrac{5}{6}$
    $+\dfrac{1}{6}$

3. $3\dfrac{4}{5}$
    $+6\dfrac{3}{5}$

4. $\dfrac{2}{3}$
    $+\dfrac{1}{4}$

5. $\dfrac{3}{4}$
    $+\dfrac{2}{5}$

6. $\dfrac{7}{12}$
    $+\dfrac{2}{3}$

Helpful Hints	To subtract fractions with unlike denominators, find the least common denominator. Multiply each fraction by one to make equivalent fractions. Finally, subtract. Reduce answers to lowest terms.	**Examples:** $\dfrac{3}{5} \times \dfrac{2}{2} = \dfrac{6}{10}$ $\dfrac{5}{6} \times \dfrac{2}{2} = \dfrac{10}{12}$ $-\dfrac{1}{2} \times \dfrac{5}{5} = \dfrac{5}{10}$ $-\dfrac{1}{4} \times \dfrac{3}{3} = \dfrac{3}{12}$ $\dfrac{1}{10}$ $\dfrac{7}{12}$

S1. $\dfrac{5}{9}$
     $-\dfrac{1}{3}$

S2. $\dfrac{5}{6}$
     $-\dfrac{1}{4}$

1. $\dfrac{7}{8}$
    $-\dfrac{4}{5}$

2. $\dfrac{9}{10}$
    $-\dfrac{1}{3}$

3. $\dfrac{4}{5}$
    $-\dfrac{1}{6}$

4. $\dfrac{11}{12}$
    $-\dfrac{2}{3}$

5. $\dfrac{5}{6}$
    $-\dfrac{2}{3}$

6. $\dfrac{11}{18}$
    $-\dfrac{2}{9}$

7. $\dfrac{5}{6}$
    $-\dfrac{7}{12}$

8. $\dfrac{4}{5}$
    $-\dfrac{1}{2}$

9. $\dfrac{7}{8}$
    $-\dfrac{2}{7}$

10. $\dfrac{11}{15}$
     $-\dfrac{1}{3}$

1.
2.
3.
4.
5.
6.
7.
8.
9.
10.
Score

Problem Solving

Jamie weighed $110\dfrac{1}{4}$ pounds two months ago. If she now weighs 115 pounds, how many pounds did she gain?

Review Exercises

1.　$\dfrac{4}{5}$ $-\dfrac{1}{5}$

2.　$\dfrac{7}{8}$ $-\dfrac{1}{8}$

3.　$3\dfrac{11}{12}$ $-2\dfrac{3}{12}$

4.　7 $-2\dfrac{11}{12}$

5.　9 $-\dfrac{3}{4}$

6.　$\dfrac{5}{6}$ $-\dfrac{1}{5}$

Helpful Hints	Use what you have learned to solve the following problems. *Remember, Reduce your answers to lowest terms.

S1.　$\dfrac{3}{4}$ $-\dfrac{1}{2}$

S2.　$\dfrac{7}{8}$ $-\dfrac{1}{5}$

1.　$\dfrac{5}{6}$ $-\dfrac{1}{3}$

2.　$\dfrac{3}{4}$ $-\dfrac{3}{16}$

3.　$\dfrac{5}{8}$ $-\dfrac{1}{4}$

4.　$\dfrac{4}{5}$ $-\dfrac{1}{20}$

5.　$\dfrac{3}{4}$ $-\dfrac{1}{3}$

6.　$\dfrac{7}{9}$ $-\dfrac{1}{18}$

7.　$\dfrac{4}{5}$ $-\dfrac{1}{2}$

8.　$\dfrac{11}{15}$ $-\dfrac{1}{3}$

9.　$\dfrac{1}{4}$ $-\dfrac{1}{6}$

10.　$\dfrac{7}{8}$ $-\dfrac{1}{6}$

1. ___
2. ___
3. ___
4. ___
5. ___
6. ___
7. ___
8. ___
9. ___
10. ___

Score ___

Problem Solving	Sally wants to send holiday cards to 75 people. If cards come in boxes of 12, how many boxes does she need to buy? How many cards will be left over?

　　　　35

Review Exercises

1. $\dfrac{7}{8}$
$+ \dfrac{7}{8}$

2. $\dfrac{4}{5}$
$+ \dfrac{4}{15}$

3. $\dfrac{2}{3}$
$+ \dfrac{1}{5}$

4. $\dfrac{3}{8}$
$- \dfrac{1}{8}$

5. 7
$-2\dfrac{3}{5}$

6. $\dfrac{4}{5}$
$- \dfrac{2}{3}$

Helpful Hints	Use what you have learned to solve the following problems. *Remember* 1. Change improper fractions to mixed numerals. 2. Reduce all answers to lowest terms.

S1. $\dfrac{7}{8}$
$- \dfrac{1}{4}$

S2. $\dfrac{5}{8}$
$+ \dfrac{3}{4}$

1. $\dfrac{5}{9}$
$+ \dfrac{1}{3}$

2. $\dfrac{3}{4}$
$- \dfrac{2}{3}$

3. $\dfrac{14}{15}$
$- \dfrac{1}{3}$

4. $\dfrac{4}{5}$
$+ \dfrac{1}{2}$

5. $\dfrac{11}{22}$
$+ \dfrac{1}{11}$

6. $\dfrac{4}{5}$
$- \dfrac{3}{10}$

7. $\dfrac{3}{5}$
$- \dfrac{1}{10}$

8. $\dfrac{11}{25}$
$+ \dfrac{2}{5}$

9. $\dfrac{4}{7}$
$- \dfrac{3}{14}$

10. $\dfrac{11}{20}$
$+ \dfrac{3}{10}$

1.
2.
3.
4.
5.
6.
7.
8.
9.
10.
Score

Problem Solving	Rhonda bought $\dfrac{3}{4}$ pounds of white chocolate candy and $\dfrac{2}{3}$ pounds of dark chocolate candy. How many pounds of candy did she by altogether?

Review Exercises

1. $3\frac{3}{8}$
$+4\frac{1}{8}$

2. $5\frac{4}{5}$
$+6\frac{3}{5}$

3. $5\frac{7}{10}$
$+4\frac{9}{10}$

4. $6\frac{3}{4}$
$-2\frac{1}{4}$

5. $7\frac{1}{10}$
$-3\frac{7}{10}$

6. 9
$-\frac{12}{15}$

Helpful Hints	Use what you have learned to solve the following problems. * Be careful to find the **lowest** common denominator. * Reduce answers to **lowest** terms.

S1. $\frac{7}{9}$ $-\frac{5}{18}$

S2. $\frac{5}{6}$ $+\frac{1}{4}$

1. $\frac{11}{12}$ $-\frac{1}{6}$

2. $\frac{11}{20}$ $+\frac{1}{5}$

3. $\frac{2}{15}$ $+\frac{3}{10}$

4. $\frac{9}{16}$ $-\frac{1}{2}$

5. $\frac{7}{8}$ $-\frac{1}{6}$

6. $\frac{2}{5}$ $+\frac{1}{3}$

7. $\frac{5}{8}$ $-\frac{1}{2}$

8. $\frac{3}{8}$ $+\frac{1}{2}$

9. $\frac{21}{25}$ $-\frac{7}{50}$

10. $\frac{8}{9}$ $+\frac{1}{3}$

1.
2.
3.
4.
5.
6.
7.
8.
9.
10.

Score

Problem Solving   Ricardo worked 8 hours on Monday. He spent $3\frac{3}{4}$ hours working on the computer. How much time did he spend working on other activities?

Review Exercises

1. Change $3\frac{2}{7}$ to an improper fraction.

2. Change $\frac{17}{15}$ to a mixed numeral.

3. Reduce $\frac{16}{28}$ to its lowest terms.

4. Shade the figure showing $\frac{3}{4}$.

5. $\begin{array}{r} \frac{4}{8} \\ + \frac{5}{8} \\ \hline \end{array}$

6. $\begin{array}{r} \frac{3}{4} \\ + \frac{3}{8} \\ \hline \end{array}$

Helpful Hints	When adding mixed numerals with unlike denominators, first add the fractions. If there is an improper fraction, make it a mixed numeral. Finally, add the sum to the sum of the whole numbers. * Reduce fractions to lowest terms.	**Example:** $3\frac{2}{3} \times \frac{2}{2} = \frac{4}{6}$ $+ \, 2\frac{1}{2} \times \frac{3}{3} = \frac{3}{6}$ $\overline{5\frac{7}{6} = 1\frac{1}{6} = 6\frac{1}{6}}$

S1. $\begin{array}{r} 3\frac{1}{3} \\ + 4\frac{2}{5} \\ \hline \end{array}$

S2. $\begin{array}{r} 5\frac{3}{4} \\ + 2\frac{1}{3} \\ \hline \end{array}$

1. $\begin{array}{r} 3\frac{3}{4} \\ + 1\frac{1}{2} \\ \hline \end{array}$

2. $\begin{array}{r} 5\frac{2}{3} \\ + 2\frac{1}{6} \\ \hline \end{array}$

3. $\begin{array}{r} 4\frac{1}{5} \\ + 3\frac{7}{10} \\ \hline \end{array}$

4. $\begin{array}{r} 3\frac{7}{10} \\ + 4\frac{1}{2} \\ \hline \end{array}$

5. $\begin{array}{r} 7\frac{2}{3} \\ + 3\frac{5}{9} \\ \hline \end{array}$

6. $\begin{array}{r} 2\frac{3}{5} \\ + 4\frac{1}{2} \\ \hline \end{array}$

7. $\begin{array}{r} 5\frac{3}{7} \\ + 2\frac{11}{14} \\ \hline \end{array}$

8. $\begin{array}{r} 6\frac{11}{18} \\ + 4\frac{4}{9} \\ \hline \end{array}$

9. $\begin{array}{r} 5\frac{1}{3} \\ + 2\frac{1}{7} \\ \hline \end{array}$

10. $\begin{array}{r} 7\frac{5}{6} \\ + 3\frac{3}{8} \\ \hline \end{array}$

1.
2.
3.
4.
5.
6.
7.
8.
9.
10.
Score

Problem Solving

It rained $7\frac{3}{4}$ inches in January. In February it rained $9\frac{1}{2}$ inches. What was the total rainfall for both months?

Review Exercises

1. $\frac{3}{4}$
$+ \frac{1}{3}$

2. $\frac{5}{7}$
$+ \frac{1}{3}$

3. $\frac{7}{16}$
$+ \frac{7}{8}$

4. $\frac{9}{10}$
$- \frac{3}{5}$

5. $\frac{11}{15}$
$- \frac{7}{30}$

6. $\frac{5}{8}$
$- \frac{1}{8}$

Helpful Hints

Use what you have learned to solve the following problems.
* Change improper fractions to mixed numerals.
* Reduce all answers to **lowest** terms.

S1. $5\frac{1}{2}$
$+ 3\frac{3}{4}$

S2. $4\frac{11}{15}$
$+ 3\frac{13}{30}$

1. $7\frac{9}{10}$
$+ 3\frac{7}{20}$

2. $6\frac{4}{5}$
$+ 2\frac{1}{10}$

3. $7\frac{5}{6}$
$+ 2\frac{1}{3}$

4. $9\frac{4}{7}$
$+ 3\frac{5}{14}$

5. $2\frac{3}{4}$
$+ 2\frac{3}{8}$

6. $7\frac{1}{2}$
$+ 8\frac{2}{3}$

7. $4\frac{3}{4}$
$+ 2\frac{2}{5}$

8. $6\frac{4}{5}$
$+ 7\frac{3}{10}$

9. $5\frac{2}{3}$
$+ \frac{1}{9}$

10. $4\frac{5}{8}$
$+ 3\frac{13}{32}$

1.
2.
3.
4.
5.
6.
7.
8.
9.
10.

Score

Problem Solving

Manuel had 20 dollars. He spent $7\frac{1}{2}$ dollars on school supplies and $3\frac{1}{4}$ dollars to watch a movie. How much money did he have left?

Review Exercises

1. $\dfrac{7}{8}$ $-\dfrac{1}{8}$

2. 7 $-2\dfrac{7}{16}$

3. $5\dfrac{3}{4}$ $-2\dfrac{1}{4}$

4. $6\dfrac{1}{3}$ $-2\dfrac{2}{3}$

5. $9\dfrac{1}{9}$ $-6\dfrac{7}{9}$

6. $\dfrac{5}{6}$ $-\dfrac{1}{3}$

Helpful Hints

To subtract mixed numerals with unlike denominators, first subtract the fractions. If the fractions cannot be subtracted, take one from the whole number, increase the fraction, then subtract.

Examples:

$$5\ \cancel{6}\,\dfrac{1}{6} = \dfrac{2}{12} + \dfrac{12}{12} = \dfrac{14}{12}$$
$$-\ 3\dfrac{1}{4} = \dfrac{3}{12}$$
$$2\dfrac{11}{12}$$

$$7\dfrac{1}{2} \times \dfrac{3}{3} = \dfrac{3}{6}$$
$$-\ 2\dfrac{1}{3} \times \dfrac{2}{2} = \dfrac{2}{6}$$
$$5 \qquad \dfrac{1}{6} = 5\dfrac{1}{6}$$

S1. $5\dfrac{1}{2}$ $-2\dfrac{1}{3}$

S2. $6\dfrac{1}{3}$ $-2\dfrac{1}{2}$

1. $7\dfrac{3}{4}$ $-2\dfrac{1}{2}$

2. $9\dfrac{4}{5}$ $-3\dfrac{1}{10}$

3. $6\dfrac{1}{5}$ $-2\dfrac{2}{3}$

4. $5\dfrac{2}{5}$ $-3\dfrac{7}{10}$

5. $6\dfrac{1}{7}$ $-2\dfrac{5}{14}$

6. $9\dfrac{1}{5}$ $-3\dfrac{9}{10}$

7. $5\dfrac{1}{3}$ $-2\dfrac{8}{9}$

8. $6\dfrac{7}{8}$ $-3\dfrac{5}{16}$

9. $3\dfrac{1}{6}$ $-1\dfrac{1}{4}$

10. $5\dfrac{1}{5}$ $-2\dfrac{7}{15}$

1.
2.
3.
4.
5.
6.
7.
8.
9.
10.
Score

Problem Solving

Paula needs 30 dollars for a new dress. She earned $8\dfrac{1}{2}$ dollars last week and another $5\dfrac{1}{4}$ dollars this week. How much more does she need in order to buy the dress?

Review Exercises

1. $3\frac{2}{5}$
$+2\frac{1}{5}$

2. $5\frac{3}{4}$
$+7\frac{3}{4}$

3. $6\frac{1}{8}$
$+3\frac{5}{8}$

4. $7\frac{1}{3}$
$+4\frac{3}{5}$

5. $8\frac{15}{16}$
$+2\frac{1}{8}$

6. $\frac{1}{2}$
$\frac{1}{3}$
$+\frac{1}{5}$

Helpful Hints	Use what you have learned to solve the following problems. * Take one from the whole number when necessary. * Reduce all answers to lowest terms.

S1. $6\frac{1}{4}$
$-2\frac{4}{8}$

S2. $5\frac{1}{4}$
$-2\frac{9}{20}$

1. $6\frac{1}{2}$
$-2\frac{1}{3}$

2. $7\frac{1}{3}$
$-2\frac{1}{2}$

3. $5\frac{2}{9}$
$-2\frac{1}{3}$

4. $4\frac{1}{6}$
$-1\frac{1}{3}$

5. $7\frac{7}{8}$
$-2\frac{7}{16}$

6. $5\frac{3}{10}$
$-1\frac{17}{30}$

7. $11\frac{1}{5}$
$-2\frac{9}{10}$

8. $5\frac{3}{10}$
$-2\frac{7}{40}$

9. $8\frac{1}{2}$
$-3\frac{3}{5}$

10. $9\frac{1}{8}$
$-2\frac{2}{3}$

1.

2.

3.

4.

5.

6.

7.

8.

9.

10.

Score

Problem Solving	Victoria started with 20 dollars. She spent $7\frac{1}{4}$ dollars, then she earned $9\frac{1}{2}$ dollars. How much money does she have now?

Review Exercises

1. $\frac{1}{2}$
$+\ \frac{1}{3}$

2. $\frac{3}{4}$
$-\ \frac{1}{8}$

3. $7\frac{1}{8}$
$-4\frac{3}{8}$

4. $7\frac{2}{3}$
$-3\frac{2}{3}$

5. $9\frac{1}{2}$
$-2\frac{2}{3}$

6. $7\frac{3}{8}$
$+3\frac{3}{4}$

Helpful Hints	Use what you have learned to solve the following problems. *Be sure all fractions are reduced to lowest terms.

S1. $3\frac{1}{5}$
$-2\frac{2}{5}$

S2. $5\frac{1}{3}$
$-2\frac{5}{6}$

1. $\frac{7}{8}$
$-\ \frac{1}{3}$

2. $\frac{8}{9}$
$-\ \frac{1}{6}$

3. 6
$-2\frac{3}{7}$

4. $6\frac{1}{8}$
$+7\frac{3}{4}$

5. $9\frac{7}{10}$
$+8\frac{9}{10}$

6. $7\frac{1}{3}$
$-2\frac{5}{9}$

7. $5\frac{3}{4}$
$-2\frac{2}{5}$

8. $9\frac{1}{2}$
-6

9. $\frac{8}{15}$
$+\ \frac{7}{30}$

10. $6\frac{2}{3}$
$+3\frac{5}{6}$

1.
2.
3.
4.
5.
6.
7.
8.
9.
10.
Score

Problem Solving	Juanito worked $7\frac{1}{2}$ hours on Monday and $9\frac{1}{2}$ hours on Tuesday. If she was paid 12 dollars per hour, how much did she earn?

Review Exercises

1. Write 2 fractions for the shaded part.

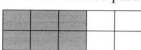

2. Change $7\frac{1}{2}$ to an improper fraction.

3. Find the missing numerator. $\frac{7}{8} = \frac{}{24}$

4. Change $\frac{31}{5}$ to a mixed numeral.

5. Reduce $\frac{40}{50}$ to its lowest terms.

6.
$$\frac{3}{5}$$
$$\frac{4}{5}$$
$$+\ \frac{2}{5}$$
_____

Helpful Hints

Use what you have learned to solve the following problems.
* If a particular problem gives you difficulty, it can be helpful to refer
to the "Helpful Hints" sections from previous lessons.

S1. $9\frac{1}{2}$
$-\ 7$
_____

S2. 7
$-\ 1\frac{3}{4}$
_____

1. $\frac{3}{4}$
$+\ \frac{1}{8}$
_____

2. $\frac{9}{16}$
$+\ \frac{11}{16}$
_____

3. $\frac{7}{8}$
$-\ \frac{1}{6}$
_____

4. $6\frac{1}{4}$
$-\ 3\frac{3}{4}$
_____

5. $5\frac{7}{8}$
$+\ 6\frac{1}{8}$
_____

6. $9\frac{5}{12}$
$-\ 2\frac{5}{6}$
_____

7. $8\frac{4}{7}$
$+\ 3\frac{9}{14}$
_____

8. $\frac{2}{3}$
$\frac{1}{4}$
$+\ \frac{1}{5}$
_____

9. $\frac{11}{12}$
$+\ \frac{5}{24}$
_____

10. $5\frac{4}{5}$
$+\ 2\frac{1}{2}$
_____

1.
_____
2.
_____
3.
_____
4.
_____
5.
_____
6.
_____
7.
_____
8.
_____
9.
_____
10.
_____
Score

Problem Solving

It takes a man $15\frac{1}{4}$ to drive to work and $18\frac{1}{2}$ minutes to drive home. What is his total commute time?

Review Exercises

1. $3\dfrac{1}{2}$
    $+2\dfrac{1}{3}$

2. $7\dfrac{1}{2}$
    $-3\dfrac{2}{3}$

3. $\dfrac{9}{10}$
    $+\dfrac{2}{5}$

4. 7
    $-2\dfrac{15}{16}$

5. $9\dfrac{11}{12}$
    $-2\dfrac{1}{12}$

6. $\dfrac{3}{4}$
    $-\dfrac{1}{12}$

Helpful Hints

When multiplying common fractions, first multiply the numerators. Next, multiply the denominators. If the answer is an improper fraction, change it to a mixed numeral.

Examples: $\dfrac{3}{4} \times \dfrac{2}{7} = \dfrac{6}{28} = \dfrac{3}{14}$ * Be sure to reduce fractions to lowest terms.

$\dfrac{3}{2} \times \dfrac{7}{8} = \dfrac{21}{16} = 1\dfrac{5}{16}$

S1. $\dfrac{3}{4} \times \dfrac{5}{7} =$

S2. $\dfrac{5}{6} \times \dfrac{2}{3} =$

1. $\dfrac{5}{9} \times \dfrac{1}{7} =$

2. $\dfrac{2}{5} \times \dfrac{1}{2} =$

3. $\dfrac{5}{2} \times \dfrac{3}{4} =$

4. $\dfrac{5}{9} \times \dfrac{2}{3} =$

5. $\dfrac{6}{5} \times \dfrac{8}{9} =$

6. $\dfrac{4}{3} \times \dfrac{4}{5} =$

7. $\dfrac{5}{2} \times \dfrac{9}{10} =$

8. $\dfrac{8}{7} \times \dfrac{4}{5} =$

9. $\dfrac{1}{2} \times \dfrac{4}{5} =$

10. $\dfrac{9}{2} \times \dfrac{3}{7} =$

1.
2.
3.
4.
5.
6.
7.
8.
9.
10.
Score

Problem Solving

Susan picked three boxes of apples. The weights were $12\dfrac{3}{4}$ pounds, $13\dfrac{1}{2}$ pounds, and $11\dfrac{1}{4}$ pounds. What was the total weight of the apples?

Review Exercises

1. Change $2\frac{3}{8}$ to an improper fraction.

3. Change $\frac{9}{2}$ to a mixed numeral.

6. $\begin{array}{r} \frac{3}{5} \\ \frac{4}{5} \\ + \frac{2}{5} \\ \hline \end{array}$

5. $\begin{array}{r} 3\frac{2}{5} \\ -2 \\ \hline \end{array}$

2. $\frac{3}{4} \times \frac{2}{3} =$

4. $\frac{2}{3} + \frac{1}{5} =$

Helpful Hints	Use what you have learned to solve the following problems. When multiplying fractions, "of" means to multiply. **Example:** Find $\frac{1}{2}$ of $\frac{3}{4}$ means: $\frac{1}{2} \times \frac{3}{4}$ * Be sure to reduce fractions to lowest terms.

S1. Find $\frac{3}{4}$ of $\frac{4}{5} =$

S2. $\frac{7}{2} \times \frac{3}{4} =$

1. Find $\frac{1}{3}$ of $\frac{8}{5} =$

2. $\frac{5}{3} \times \frac{3}{4} =$

3. $\frac{9}{10} \times \frac{3}{2} =$

4. $\frac{1}{2}$ of $\frac{8}{3} =$

5. $\frac{3}{4} \times \frac{5}{6} =$

6. $\frac{5}{4} \times \frac{3}{2} =$

7. $\frac{7}{3} \times \frac{3}{2} =$

8. Find $\frac{2}{3}$ of $\frac{5}{6} =$

9. $\frac{5}{6} \times \frac{3}{2} =$

10. $\frac{3}{4} \times \frac{8}{9} =$

1.

2.

3.

4.

5.

6.

7.

8.

9.

10.

Score

Problem Solving	Julia bought a CD that cost 23 dollars. If she paid for it with two twenty dollar bills, how much change should she receive back?

Review Exercises

1. $\frac{1}{2}$ of $\frac{3}{4}$ =

3. $\frac{3}{5}$
    $+ \frac{1}{2}$
    _____

5. $7\frac{3}{4}$
    $- 4$
    _____

6. $9\frac{1}{3}$
    $-2\frac{2}{3}$
    _____

4. $5\frac{1}{2}$
    $-2\frac{1}{3}$
    _____

2. $\frac{5}{2} \times \frac{2}{3}$ =

Helpful Hints

If the numerator of one fraction and the denominator of another have a common factor, they can be divided out before you multiply the fractions.

Examples:

$\frac{3}{\underset{1}{4}} \times \frac{\overset{2}{8}}{11} = \frac{6}{11}$

4 is a common factor.

$\frac{7}{\underset{4}{8}} \times \frac{\overset{3}{6}}{5} = \frac{21}{20} = 1\frac{1}{20}$

2 is a common factor.

S1. $\frac{3}{5} \times \frac{7}{9}$ =

S2. $\frac{12}{15} \times \frac{5}{6}$ =

1. $\frac{3}{5} \times \frac{15}{16}$ =

2. $\frac{4}{15} \times \frac{5}{16}$ =

3. $\frac{5}{6}$ of $\frac{11}{15}$ =

4. $\frac{7}{3} \times \frac{11}{7}$ =

5. $\frac{5}{8} \times \frac{12}{25}$ =

6. $\frac{8}{9} \times \frac{3}{4}$ =

7. $\frac{3}{4} \times \frac{8}{15}$ =

8. $\frac{3}{4}$ of $\frac{3}{5}$ =

9. $\frac{5}{3} \times \frac{6}{7}$ =

10. $\frac{11}{6} \times \frac{4}{7}$ =

1.	
2.	
3.	
4.	
5.	
6.	
7.	
8.	
9.	
10.	
Score	

Problem Solving

There are 12 rows of seats in a theater. Each row has 15 seats. If 97 seats are occupied, how many seats are empty?

Fractions

Review Exercises

1. Change $\frac{25}{15}$ to a mixed numeral.

2. $\frac{3}{4}$ of $\frac{5}{6} =$

3. $\frac{7}{2} \times \frac{1}{3} =$

4. Change $12\frac{1}{2}$ to an improper fraction.

5. $3\frac{5}{6}$ $+2\frac{1}{3}$

6. $\frac{11}{16}$ $-\frac{1}{4}$

Helpful Hints

Use what you have learned to solve the following problems.
* Change improper fractions to mixed numerals.
* Reduce fractions to lowest terms.

S1. $\frac{5}{6} \times \frac{12}{25} =$

S2. Find $\frac{7}{8}$ of $\frac{24}{21} =$

1. $\frac{9}{10} \times \frac{7}{18} =$

2. $\frac{5}{6} \times \frac{8}{9} =$

3. $\frac{5}{6} \times \frac{11}{20} =$

4. $\frac{18}{25} \times \frac{50}{9} =$

5. $\frac{1}{2}$ of $\frac{9}{2} =$

6. $\frac{7}{6} \times \frac{3}{14} =$

7. $\frac{11}{3} \times \frac{9}{22} =$

8. $\frac{4}{3} \times \frac{15}{16} =$

9. $\frac{9}{10}$ of $\frac{40}{27} =$

10. $\frac{8}{5} \times \frac{11}{24} =$

1.

2.

3.

4.

5.

6.

7.

8.

9.

10.

Score

Problem Solving

The Gonzales family has a yearly income of 156,000 dollars. What is their average monthly income? (Hint: 1 year = 12 months).

Review Exercises

1. Reduce $\frac{40}{48}$ to lowest terms.

2. Change $\frac{27}{2}$ to a mixed numeral.

3. Change $5\frac{4}{5}$ to an improper fraction.

4. $\frac{3}{4} \times \frac{16}{17} =$

5. Find $\frac{2}{3}$ of $\frac{21}{40} =$

6. $\frac{7}{3} \times \frac{13}{14} =$

Helpful Hints

When multiplying whole numbers and fractions, write the whole number as a fraction and then multiply. See examples:

$\frac{2}{3} \times 15 =$

$\frac{2}{3} \times \frac{15}{1} = \frac{10}{1} = 10$

$\frac{3}{4} \times 9 =$

$\frac{3}{4} \times \frac{9}{1} = \frac{27}{4}$

$6\frac{3}{4}$ $4\overline{)27}$ $\frac{24}{3}$

1.
2.
3.
4.
5.
6.
7.
8.
9.
10.
Score

S1. $\frac{4}{5} \times 15 =$

S2. $\frac{3}{4} \times 5 =$

1. $\frac{3}{4} \times 16 =$

2. $12 \times \frac{5}{6} =$

3. $\frac{3}{4}$ of $24 =$

4. $\frac{3}{7} \times 5 =$

5. $\frac{1}{2} \times 37 =$

6. $\frac{1}{10} \times 15 =$

7. $6 \times \frac{7}{18} =$

8. Find $\frac{5}{6}$ of $9 =$

9. $\frac{7}{8} \times 40 =$

10. $\frac{3}{5} \times 7 =$

Problem Solving

A class has 35 students. If $\frac{2}{5}$ of them are girls, how many girls are there in the class. (Hint: "of" means to multiply).

Review Exercises

1. $\dfrac{4}{5}$
   $+\ \dfrac{4}{5}$

2. $3\dfrac{1}{2}$
   $+4\dfrac{5}{6}$

3. $7\dfrac{11}{12}$
   $+2\dfrac{1}{2}$

4. $\dfrac{11}{15}$
   $-\dfrac{1}{15}$

5. $\dfrac{11}{20}$
   $-\dfrac{3}{10}$

6. $4\dfrac{1}{8}$
   $-3\dfrac{3}{4}$

Helpful Hints	Use what you have learned to solve the following problems. * Change improper fractions to mixed numerals. * Reduce fractions to lowest terms.

S1. $\dfrac{4}{5} \times 7 =$ S2. $\dfrac{5}{8} \times 40 =$ 1. $\dfrac{4}{5}$ of $25 =$

2. $\dfrac{4}{5} \times 12 =$ 3. $\dfrac{2}{3} \times 10 =$ 4. $\dfrac{3}{4}$ of $60 =$

5. $\dfrac{1}{4} \times 15 =$ 6. $\dfrac{2}{3} \times 30 =$ 7. $55 \times \dfrac{7}{11} =$

8. $\dfrac{5}{6} \times 20 =$ 9. $12 \times \dfrac{2}{5} =$ 10. $20 \times \dfrac{1}{6} =$

1.

2.

3.

4.

5.

6.

7.

8.

9.

10.

Score

Problem Solving	A school has 300 students. $\dfrac{2}{5}$ of the students are boys. How many of the students are girls?

Review Exercises

1. $\dfrac{2}{3} \times \dfrac{5}{7} =$ 4. Find $\dfrac{2}{5}$ of $25 =$

2. $\dfrac{5}{6} \times \dfrac{12}{13} =$ 5. $7 \times \dfrac{2}{3} =$

3. $\dfrac{24}{25} \times \dfrac{50}{8} =$ 6. $\dfrac{3}{4} \times 32 =$

| **Helpful Hints** | To multiply mixed numerals, first change them to improper fractions, then multiply. Express answers in lowest terms. | **Example:** $1\dfrac{1}{2} \times 1\dfrac{5}{6} =$ $\dfrac{\cancel{3}^{1}}{2} \times \dfrac{11}{\cancel{6}_{2}} = \dfrac{11}{4} = 2\dfrac{3}{4}$ |

S1. $\dfrac{1}{3} \times 1\dfrac{1}{3} =$ S2. $2\dfrac{1}{4} \times 2\dfrac{1}{3} =$ 1. $\dfrac{1}{3} \times 3\dfrac{1}{2} =$

2. $3\dfrac{1}{3} \times 2\dfrac{1}{5} =$ 3. $4 \times 2\dfrac{3}{4} =$ 4. $3\dfrac{1}{7} \times 1\dfrac{2}{5} =$

5. $3\dfrac{2}{3} \times 2\dfrac{1}{4} =$ 6. $2\dfrac{1}{2} \times 3\dfrac{1}{4} =$ 7. $2\dfrac{1}{2} \times 3\dfrac{1}{2} =$

8. $2\dfrac{1}{2} \times 8 =$ 9. $3\dfrac{1}{2} \times 4\dfrac{2}{3} =$ 10. $3\dfrac{1}{6} \times \dfrac{6}{7} =$

| 1. |
| 2. |
| 3. |
| 4. |
| 5. |
| 6. |
| 7. |
| 8. |
| 9. |
| 10. |
| Score |

| **Problem Solving** | If a long distance runner can run 8 miles in an hour, how far can he run in $4\dfrac{1}{2}$ hours? |

Review Exercises

1. $\dfrac{7}{8} \times \dfrac{4}{9} =$

2. $5 \times \dfrac{3}{4} =$

3. $\dfrac{2}{3} \times 24 =$

4. $\begin{array}{r} 7\frac{1}{2} \\ -1\frac{1}{3} \\ \hline \end{array}$

5. $\begin{array}{r} 5\frac{1}{3} \\ -2\frac{2}{5} \\ \hline \end{array}$

6. $\begin{array}{r} 3\frac{2}{5} \\ +4\frac{7}{10} \\ \hline \end{array}$

Helpful Hints	Use what you have learned to solve the following problems. * Divide out common factors. * Reduce fractions to lowest terms.

S1. $3\frac{1}{2} \times 1\frac{1}{7} =$ S2. $2\frac{2}{3} \times 3\frac{3}{4} =$ 1. $\dfrac{9}{10} \times 2\frac{1}{2} =$

2. $3\frac{1}{8} \times 2\frac{1}{5} =$ 3. $3\frac{1}{2} \times 2\frac{1}{2} =$ 4. $9 \times 2\frac{2}{3} =$

5. $5\frac{1}{3} \times 2\frac{3}{4} =$ 6. $4\frac{1}{2} \times \dfrac{3}{4} =$ 7. $5\frac{1}{4} \times 2\frac{1}{7} =$

8. $6\frac{1}{2} \times 1\frac{1}{3} =$ 9. $3\frac{1}{2} \times 2\frac{1}{3} =$ 10. $6\frac{1}{4} \times 1\frac{3}{5} =$

1.

2.

3.

4.

5.

6.

7.

8.

9.

10.

Score

Problem Solving	If a factory can produce 50 cars per day, how many cars can be produced in $5\frac{1}{2}$ days?

Review Exercises

1. $\dfrac{2}{3} \times \dfrac{7}{8} =$ 2. $\dfrac{3}{4} \times 16 =$ 3. $\dfrac{2}{3} \times 2\dfrac{1}{2} =$

4. $5 \times 3\dfrac{1}{2} =$ 5. $2\dfrac{1}{2} \times 1\dfrac{1}{5} =$ 6. $3\dfrac{1}{3} \times 1\dfrac{1}{10} =$

Helpful Hints	Use what you have learned to solve the following problems. * Be sure to express answers in lowest terms. * Sometimes common factors may be divided out before you multiply.

1.
2.
3.
4.
5.
6.
7.
8.
9.
10.
Score

S1. $\dfrac{7}{8} \times \dfrac{4}{9} =$ S2. $3\dfrac{1}{2} \times 1\dfrac{1}{7} =$ 1. $\dfrac{4}{5} \times \dfrac{3}{8} =$

2. $\dfrac{9}{4} \times \dfrac{8}{11} =$ 3. $\dfrac{3}{5} \times 45 =$ 4. $\dfrac{3}{8} \times 5 =$

5. $\dfrac{3}{5} \times 3\dfrac{1}{2} =$ 6. $2\dfrac{2}{3} \times \dfrac{1}{3} =$ 7. $4 \times 2\dfrac{3}{4} =$

8. $1\dfrac{3}{5} \times 1\dfrac{1}{3} =$ 9. $4\dfrac{1}{6} \times 4\dfrac{4}{5} =$ 10. $9 \times 1\dfrac{2}{5} =$

Problem Solving	Chuck has 240 dollars in his savings account. If he spends $\dfrac{3}{4}$ of his savings, how much does he have left in his account?

Review Exercises

1. $\dfrac{3}{4}$
    $-\dfrac{1}{2}$

2. $\dfrac{7}{8}$
    $-\dfrac{1}{3}$

3. 7
    $-2\dfrac{14}{15}$

4. $7\dfrac{1}{5}$
    $-2\dfrac{4}{5}$

5. $9\dfrac{5}{6}$
    $-1\dfrac{1}{4}$

6. $3\dfrac{2}{5}$
    $-1\dfrac{7}{10}$

Helpful Hints

Use what you have learned to solve the following problems.
* If you have difficulty, refer to previous "Helpful Hints" sections for help.

S1. $\dfrac{3}{4}$ of $3\dfrac{1}{3} =$

S2. $2\dfrac{1}{4} \times 3\dfrac{1}{2} =$

1. $\dfrac{15}{16} \times \dfrac{8}{30} =$

2. $\dfrac{7}{2} \times \dfrac{9}{14} =$

3. $15 \times \dfrac{3}{5} =$

4. $\dfrac{1}{5} \times 4\dfrac{1}{2} =$

5. $3\dfrac{3}{4} \times 1\dfrac{4}{5} =$

6. $60 \times 2\dfrac{1}{2} =$

7. $3\dfrac{2}{3} \times 1\dfrac{5}{11} =$

8. $6 \times 3\dfrac{3}{4} =$

9. $2\dfrac{1}{4} \times 2\dfrac{1}{3} =$

10. $6 \times 3\dfrac{1}{2} =$

1.	
2.	
3.	
4.	
5.	
6.	
7.	
8.	
9.	
10.	
Score	

Problem Solving

In a class of 36 students, $\dfrac{2}{3}$ of them received A's on a test.
How many students received A's.

Review Exercises

1. $\dfrac{3}{4}$ of $1\dfrac{1}{5}$ =

2. $\dfrac{25}{21} \times \dfrac{7}{30}$ =

3. $5\dfrac{3}{8}$
$+ 3\dfrac{3}{4}$
_____

5. $3\dfrac{1}{5}$
$- 1\dfrac{3}{5}$
_____

4. $2\dfrac{1}{2} \times 5$ =

6. $2\dfrac{1}{2} \times 1\dfrac{4}{5}$ =

Helpful Hints	To find the reciprocal of a common fraction, invert the fraction. To find the reciprocal of a mixed numeral, change the mixed numeral to an improper fraction, then invert it. To find the reciprocal of a whole number first make it a fraction then invert it.	**Examples:** The reciprocal of: $2\dfrac{1}{2}$ $\left(\dfrac{5}{2}\right)$ is $\dfrac{2}{5}$ $\dfrac{3}{5}$ is $\dfrac{5}{3}$ or $1\dfrac{2}{3}$ $\quad$ $7\left(\dfrac{7}{1}\right)$ is $\dfrac{1}{7}$

Find the reciprocal of each number.

S1. $\dfrac{2}{3}$

S2. $3\dfrac{1}{2}$

1. 7

2. $\dfrac{5}{8}$

3. $4\dfrac{1}{4}$

4. 15

5. $\dfrac{2}{7}$

6. $\dfrac{1}{9}$

7. 12

8. $\dfrac{2}{9}$

9. $5\dfrac{1}{2}$

10. 17

1.	
2.	
3.	
4.	
5.	
6.	
7.	
8.	
9.	
10.	
Score	

Problem Solving	Five students earned a total of 375 dollars. If they divided the money equally among themselves, how much did each student receive?

Fractions

Review Exercises

1. Change $7\frac{1}{2}$ to an improper fraction.

2. Reduce $\frac{18}{40}$ to its lowest terms.

3. Change $\frac{16}{7}$ to a mixed numeral.

4. Find the missing numerator.
$$\frac{7}{8} = \frac{}{56}$$

5. Shade in the fraction showing $\frac{3}{4}$.

6. Write 2 fractions for the shaded part.

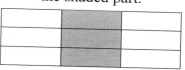

Helpful Hints

Use what you have learned to solve the following problems

Find the reciprocal of each number.

S1. $\frac{7}{8}$

S2. $4\frac{3}{4}$

1. $\frac{11}{15}$

2. 6

3. $3\frac{1}{9}$

4. 16

5. $\frac{2}{15}$

6. $\frac{1}{6}$

7. 50

8. $\frac{3}{12}$

9. $7\frac{2}{3}$

10. $\frac{11}{16}$

1.

2.

3.

4.

5.

6.

7.

8.

9.

10.

Score

Problem Solving

If the normal temperature for an adult is $98\frac{3}{4}$ degrees, and a man has a temperature of 101 degrees, how much above normal is his temperature?

Review Exercises

1. $\dfrac{3}{4}$
   $+ \dfrac{2}{3}$

2. $\dfrac{3}{5}$ of $2\dfrac{1}{2} =$

3. $7 - 2\dfrac{1}{3} =$

4. $3\dfrac{1}{2} \times 1\dfrac{4}{7} =$

5. $\dfrac{2}{3}$
   $\dfrac{1}{2}$
   $+ \dfrac{1}{4}$

6. $3\dfrac{1}{8}$
   $- 1\dfrac{7}{8}$

Helpful Hints

To divide fractions, find the reciprocal of the second number, then multiply the fractions. **Examples:**

$\dfrac{2}{3} \div \dfrac{1}{2} =$

$\dfrac{2}{3} \times \dfrac{2}{1} = \dfrac{4}{3} = \boxed{1\dfrac{1}{3}}$

$2\dfrac{1}{2} \div 1\dfrac{1}{2} = \dfrac{5}{2} \div \dfrac{3}{2}$

$\dfrac{5}{\underset{1}{2}} \times \dfrac{\overset{1}{2}}{3} = \dfrac{5}{3} = \boxed{1\dfrac{2}{3}}$

S1. $\dfrac{3}{5} \div \dfrac{3}{8} =$

S2. $4\dfrac{1}{2} \div 2 =$

1. $\dfrac{5}{8} \div \dfrac{1}{6} =$

2. $\dfrac{1}{4} \div \dfrac{1}{3} =$

3. $7 \div \dfrac{2}{3} =$

4. $5\dfrac{1}{2} \div \dfrac{1}{2} =$

5. $1\dfrac{3}{4} \div \dfrac{3}{8} =$

6. $3\dfrac{1}{4} \div \dfrac{11}{12} =$

7. $1\dfrac{1}{3} \div 3 =$

8. $7\dfrac{1}{2} \div 2 =$

9. $7\dfrac{1}{2} \div 2\dfrac{1}{2} =$

10. $4\dfrac{2}{3} \div 2\dfrac{1}{2} =$

1.
2.
3.
4.
5.
6.
7.
8.
9.
10.
Score

Problem Solving

$4\dfrac{1}{2}$ yards if cloth is to be divided into pieces that are $\dfrac{1}{2}$ yards long. How many pieces will there be?

Review Exercises

1. $\dfrac{3}{4} \times \dfrac{7}{9} =$ 2. $\dfrac{3}{4} \div \dfrac{1}{4} =$ 3. $2\dfrac{1}{2} \times 2 =$

4. $2\dfrac{1}{2} \div 2 =$ 5. $1\dfrac{1}{3} \times \dfrac{3}{4} =$ 6. $1\dfrac{1}{3} \div \dfrac{3}{4} =$

Helpful Hints	Use what you have learned to solve the following problems. * Remember, find the reciprocal of the second number, then multiply.

S1. $\dfrac{7}{8} \div \dfrac{3}{4} =$ S2. $2\dfrac{1}{2} \div 1\dfrac{1}{4} =$ 1. $\dfrac{5}{6} \div \dfrac{1}{3} =$

1.	
2.	

2. $\dfrac{3}{7} \div \dfrac{1}{2} =$ 3. $5 \div 1\dfrac{1}{2} =$ 4. $6\dfrac{1}{2} \div \dfrac{1}{4} =$

3.	
4.	
5.	

5. $1\dfrac{5}{6} \div \dfrac{1}{2} =$ 6. $4\dfrac{1}{2} \div 1\dfrac{1}{3} =$ 7. $2\dfrac{2}{3} \div \dfrac{1}{3} =$

6.	
7.	
8.	

8. $3\dfrac{3}{4} \div 2 =$ 9. $4\dfrac{1}{2} \div 2\dfrac{2}{3} =$ 10. $12\dfrac{1}{2} \div 2\dfrac{1}{2} =$

9.	
10.	
Score	

Problem Solving	A floor tile is $\dfrac{3}{4}$ inches thick. How many inches thick would a stack of 36 tiles be?

Review Exercises

1. $\dfrac{3}{8}$

    $+\dfrac{3}{4}$

2. $\dfrac{7}{16}$

    $-\dfrac{1}{4}$

3. $\dfrac{3}{4}$ of $1\dfrac{1}{2} =$

4. $\dfrac{7}{8} \div \dfrac{1}{4} =$

5. $3\dfrac{1}{2} \div \dfrac{3}{4} =$

6. $2\dfrac{2}{3} \div 1\dfrac{1}{3} =$

Helpful Hints	Sometimes it is necessary to compare the values of fractions. **Example:** $\dfrac{3}{4}$ and $\dfrac{2}{3}$ LCD is 12.

1. Find the least common denominator (LCD).
2. Change each fraction to an equivalent fraction and compare.

$\dfrac{3}{4} \times \dfrac{3}{3} = \boxed{\dfrac{9}{12}}$ $\dfrac{2}{3} \times \dfrac{4}{4} = \boxed{\dfrac{8}{12}}$ $\dfrac{3}{4}$ is the larger fraction.

Find the larger of each pair of fractions.

S1. $\dfrac{5}{8}$, $\dfrac{6}{7}$

S2. $\dfrac{7}{9}$, $\dfrac{5}{6}$

1. $\dfrac{11}{12}$, $\dfrac{8}{9}$

2. $\dfrac{3}{5}$, $\dfrac{4}{7}$

3. $\dfrac{7}{10}$, $\dfrac{5}{6}$

4. $\dfrac{9}{11}$, $\dfrac{5}{8}$

5. $\dfrac{5}{6}$, $\dfrac{13}{15}$

6. $\dfrac{2}{7}$, $\dfrac{3}{8}$

7. $\dfrac{11}{15}$, $\dfrac{23}{30}$

8. $\dfrac{1}{6}$, $\dfrac{2}{11}$

9. $\dfrac{9}{10}$, $\dfrac{8}{9}$

10. $\dfrac{11}{12}$, $\dfrac{7}{8}$

1.
2.
3.
4.
5.
6.
7.
8.
9.
10.
Score

Problem Solving	A group of hikers need to travel 55 miles. They hike 6 miles per day. After 6 days, how much farther did they still have to hike?

Review Exercises

1. $\dfrac{3}{4} \div \dfrac{1}{2} =$

2. $2\dfrac{1}{2} \div \dfrac{1}{4} =$

3. $3\dfrac{3}{4} \div 1\dfrac{1}{4} =$

4. $\dfrac{3}{5}$ of $1\dfrac{1}{2} =$

5. $5 \times 1\dfrac{3}{8} =$

6. $7\dfrac{1}{2} \times 1\dfrac{2}{5} =$

Helpful Hints

Use what you have learned to solve the following problems.

Find the larger of each pair of fractions.

S1. $\dfrac{3}{4}$, $\dfrac{7}{9}$	1.
S2. $\dfrac{11}{20}$, $\dfrac{2}{5}$	2.
1. $\dfrac{11}{14}$, $\dfrac{3}{7}$	3.

S1. $\dfrac{3}{4}$, $\dfrac{7}{9}$ S2. $\dfrac{11}{20}$, $\dfrac{2}{5}$ 1. $\dfrac{11}{14}$, $\dfrac{3}{7}$

2. $\dfrac{10}{12}$, $\dfrac{4}{5}$ 3. $\dfrac{3}{4}$, $\dfrac{4}{7}$ 4. $\dfrac{1}{2}$, $\dfrac{5}{11}$

5. $\dfrac{2}{6}$, $\dfrac{3}{11}$ 6. $\dfrac{11}{20}$, $\dfrac{3}{10}$ 7. $\dfrac{5}{6}$, $\dfrac{3}{4}$

8. $\dfrac{5}{6}$, $\dfrac{7}{8}$ 9. $\dfrac{5}{9}$, $\dfrac{3}{5}$ 10. $\dfrac{2}{7}$, $\dfrac{3}{8}$

1.

2.

3.

4.

5.

6.

7.

8.

9.

10.

Score

Problem Solving

Ruth was born in 1989. How old will she be in 2012?

Fraction Concepts - Final Review

Reduce all answers to lowest terms.

1. Reduce $\frac{20}{25}$ to its lowest terms.

2. Reduce $\frac{24}{30}$ to its lowest terms.

3. Reduce $\frac{40}{48}$ to its lowest terms.

4. Change $\frac{11}{5}$ to a mixed numeral.

5. Change $\frac{28}{12}$ to a mixed numeral.

6. Change $\frac{87}{25}$ to a mixed numeral.

7. Change $5\frac{3}{5}$ to an improper fraction.

8. Change $12\frac{4}{5}$ to an improper fraction.

9. Change $7\frac{1}{3}$ to an improper fraction.

10. Find the missing numerator.
$$\frac{5}{6} = \frac{}{48}$$

11. Find the missing numerator.
$$\frac{}{49} = \frac{6}{7}$$

12. Find the missing numerator.
$$\frac{5}{6} = \frac{}{72}$$

13. Find the LCD. $\frac{1}{6}$, $\frac{1}{4}$, $\frac{1}{3}$

14. Find the LCD. $\frac{1}{5}$, $\frac{1}{3}$, $\frac{3}{4}$

15. Find the LCD. $\frac{5}{6}$, $\frac{1}{4}$, $\frac{1}{2}$

16. Write 2 fractions for the shaded part.

17. Write 2 fractions for the shaded part.

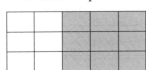

18. Shade in $\frac{3}{5}$ of the figure.

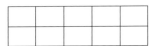

19. Find the larger fraction.
$$\frac{5}{8} , \frac{2}{3}$$

20. Find the larger fraction.
$$\frac{5}{6} , \frac{7}{9}$$

1.
2.
3.
4.
5.
6.
7.
8.
9.
10.
11.
12.
13.
14.
15.
16.
17.
18.
19.
20.

Fraction Operations - Final Review

Reduce all answers to lowest terms.

1. $\dfrac{3}{7}$
    $+ \dfrac{4}{7}$

2. $\dfrac{5}{8}$
    $+ \dfrac{7}{8}$

3. $\dfrac{3}{4}$
    $+ \dfrac{2}{5}$

4. $3\dfrac{3}{4}$
    $+2\dfrac{5}{8}$

5. $5\dfrac{2}{3}$
    $+3\dfrac{5}{6}$

6. $\dfrac{7}{8}$
    $- \dfrac{1}{8}$

7. $5\dfrac{1}{7}$
    $-2\dfrac{3}{7}$

8. 7
    $-2\dfrac{4}{7}$

9. $6\dfrac{3}{4}$
    $-2\dfrac{1}{2}$

10. $7\dfrac{1}{3}$
    $-2\dfrac{4}{5}$

11. $\dfrac{3}{4} \times \dfrac{2}{7} =$

12. $\dfrac{15}{16} \times \dfrac{7}{30} =$

13. $\dfrac{4}{5} \times 30 =$

14. $\dfrac{3}{4} \times 4\dfrac{2}{3} =$

15. $2\dfrac{1}{2} \times 3\dfrac{1}{2} =$

16. $\dfrac{5}{6} \div \dfrac{1}{3} =$

17. $3\dfrac{1}{2} \div \dfrac{1}{4} =$

18. $3\dfrac{3}{4} \div 2\dfrac{1}{2} =$

19. $3\dfrac{3}{4} \div 1\dfrac{3}{8} =$

20. $5 \div 2\dfrac{1}{5} =$

Answers
1.
2.
3.
4.
5.
6.
7.
8.
9.
10.
11.
12.
13.
14.
15.
16.
17.
18.
19.
20.

Fraction Concepts - Final Test

Reduce all answers to lowest terms.

1. Reduce $\dfrac{24}{28}$ to its lowest terms.

2. Reduce $\dfrac{18}{30}$ to its lowest terms.

3. Reduce $\dfrac{80}{100}$ to its lowest terms.

4. Change $\dfrac{13}{4}$ to a mixed numeral.

5. Change $\dfrac{19}{12}$ to a mixed numeral.

6. Change $\dfrac{20}{14}$ to a mixed numeral.

7. Change $2\dfrac{1}{8}$ to an improper fraction.

8. Change $8\dfrac{2}{3}$ to an improper fraction.

9. Change $2\dfrac{5}{6}$ to an improper fraction.

10. Find the missing numerator.
$$\dfrac{3}{12} = \dfrac{}{36}$$

11. Find the missing numerator.
$$\dfrac{}{20} = \dfrac{4}{5}$$

12. Find the missing numerator.
$$\dfrac{7}{8} = \dfrac{}{56}$$

13. Find the LCD. $\dfrac{1}{3}$, $\dfrac{1}{2}$, $\dfrac{1}{5}$

14. Find the LCD. $\dfrac{3}{8}$, $\dfrac{5}{6}$

15. Find the LCD. $\dfrac{1}{2}$, $\dfrac{1}{5}$, $\dfrac{1}{6}$

16. Write 2 fractions for the shaded part.

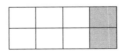

17. Write 2 fractions for the shaded part.

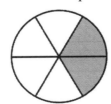

18. Shade in $\dfrac{3}{4}$ of the figure.

19. Find the larger fraction.
$$\dfrac{4}{7} \ , \ \dfrac{6}{11}$$

20. Find the larger fraction.
$$\dfrac{3}{4} \ , \ \dfrac{4}{5} \ , \ \dfrac{1}{2}$$

1.	
2.	
3.	
4.	
5.	
6.	
7.	
8.	
9.	
10.	
11.	
12.	
13.	
14.	
15.	
16.	
17.	
18.	
19.	
20.	

Fraction Operations - Final Test

Reduce all answers to lowest terms.

1. $\dfrac{3}{5}$
$+\ \dfrac{1}{5}$

2. $\dfrac{7}{12}$
$+\ \dfrac{11}{12}$

3. $\dfrac{3}{5}$
$+\ \dfrac{7}{10}$

4. $2\dfrac{3}{5}$
$+3\dfrac{7}{15}$

5. $5\dfrac{3}{7}$
$+2\dfrac{2}{3}$

6. $\dfrac{9}{10}$
$-\ \dfrac{1}{10}$

7. $6\dfrac{1}{8}$
$-2\dfrac{3}{8}$

8. 6
$-2\dfrac{3}{8}$

9. $6\dfrac{3}{4}$
$-2\dfrac{1}{5}$

10. $7\dfrac{5}{12}$
$-2\dfrac{3}{4}$

11. $\dfrac{1}{2} \times \dfrac{8}{9} =$

12. $\dfrac{8}{15} \times \dfrac{5}{16} =$

13. $\dfrac{3}{4} \times 28 =$

14. $\dfrac{4}{5} \times 3\dfrac{1}{4} =$

15. $2\dfrac{2}{5} \times 2\dfrac{1}{2} =$

16. $\dfrac{3}{8} \div \dfrac{1}{4} =$

17. $3\dfrac{2}{3} \div \dfrac{1}{2} =$

18. $4\dfrac{1}{2} \div 2\dfrac{1}{3} =$

19. $5 \div 2\dfrac{1}{8} =$

20. $4\dfrac{1}{2} \div 3 =$

| 1. |
| 2. |
| 3. |
| 4. |
| 5. |
| 6. |
| 7. |
| 8. |
| 9. |
| 10. |
| 11. |
| 12. |
| 13. |
| 14. |
| 15. |
| 16. |
| 17. |
| 18. |
| 19. |
| 20. |

PAGE 4
Review Exercises:
1. 72
2. 549
3. 1,264
4. 328
5. 448
6. 324
S1. 1/4; 3/4
S2. 5/6; 1/6
1. 6/8, 3/4; 2/8, 1/4
2. 1/3; 2/3
3. 2/4, 1/2; 2/4, 1/2
4. 3/8, 5/8
5. 5/8; 3/8
6. 2/3; 1/3
7. 5/6; 1/6
8. 1/8; 7/8
9. 1/3, 2/6; 2/3, 4/6
10. 7/8; 1/8
Problem Solving: 360 crayons

PAGE 5
Review Exercises:
1. 127
2. 629
3. 536
4. 1,498
5. 435
6. 173
S1. 1/3; 2/3
S2. 4/8, 2/4, 1/2; 4/8, 2/4, 1/2
1. 7/8; 1/8
2. 2/8, 1/4; 6/8, 3/4
3. 6/8, 3/4; 2/8, 1/4
4. 1/3; 2/3
5. 4/8, 2/4, 1/2; 4/8, 2/4, 1/2
6. 3/4; 1/4
7. 3/9, 1/3; 6/9, 2/3
8. 6/9, 2/3; 3/9, 1/3
9. 3/8; 5/8
10. 2/4, 1/2; 2/4, 1/2
Problem Solving: $771

PAGE 6
Review Exercises:
1. 1/2
2. 6/9, 2/3
3. 3/9, 1/3
4. 3/6, 1/2
5. 2/8, 1/4
6. 6/8, 3/4
S1. answers vary, one fifth
S2. answers vary, three eighths
1. answers vary, three fourths
2. answers vary, two ninths
3. answers vary, four fifths
4. answers vary, four sevenths
5. answers vary, seven twelfths
6. answers vary, seven eighths
7. answers vary, four eighths
8. answers vary, two thirds
9. answers vary, five sixths
10. answers vary, four sixths
Problem Solving: 440 miles

PAGE 7
Review Exercises:
1. 457
2. answers vary
3. 6/8, 3/4
4. 465
5. 682
6. 192
S1. answers vary, one third
S2. answers vary, three tenths
1. answers vary, five eighths
2. answers vary, one sixth
3. answers vary, one eighth
4. answers vary, three twelfths
5. answers vary, two thirds
6. answers vary, seven ninths
7. answers vary, two sevenths
8. answers vary, six ninths
9. answers vary, nine tenths
10. answers vary, two fourths
Problem Solving: 27

PAGE 8
Review Exercises:
1. 2/4, 1/2
2. 6/8, 3/4
3. answers vary
4. 792
5. 437
6. 367
S1. 1/2
S2. 3/4
1. 4/5
2. 3/4
3. 1/2
4. 4/5
5. 2/3
6. 2/3
7. 3/5
8. 5/8
9. 5/6
10. 3/4
Problem Solving: 12 gallons

PAGE 9
Review Exercises:
1. 3/4
2. 4/5
3. 89
4. answers vary
5. 1,363
6. 13,820
S1. 3/4
S2. 5/6
1. 3/5
2. 3/4
3. 8/15
4. 7/8
5. 3/4
6. 5/8
7. 15/16
8. 5/6
9. 8/9
10. 5/6
Problem Solving: 14 miles

Solutions

PAGE 10
Review Exercises:
1. 348
2. 789
3. 6,237
4. 3/4
5. 457
6. 5/6
S1. 15
S2. 9
1. 10
2. 12
3. 18
4. 22
5. 35
6. 18
7. 6
8. 40
9. 16
10. 33
Problem Solving: $30

PAGE 11
Review Exercises:
1. 3/4
2. 281
3. 4/6, 2/3
4. 171
5. 21
6. 7,285
S1. 18
S2. 30
1. 12
2. 44
3. 35
4. 30
5. 18
6. 9
7. 6
8. 33
9. 32
10. 49
Problem Solving: 39 miles per gallon

PAGE 12
Review Exercises:
1. 5/8
2. 2/3
3. 14/25
4. 235
5. 1981
6. 1/2, 2/4, 4/8
S1. 1 1/4
S2. 1 1/2
1. 2 3/4
2. 1 4/7
3. 3
4. 5 1/2
5. 3 1/12
6. 4
7. 5 1/2
8. 6 2/3
9. 2 1/4
10. 5 2/5
Problem Solving: 3,734

PAGE 13
Review Exercises:
1. 1 4/7
2. 9
3. 5/6
4. 6/9, 2/3
5. 6,262
6. 1,980
S1. 1 4/15
S2. 3 1/4
1. 2 1/2
2. 1 1/8
3. 5 2/5
4. 5 1/3
5. 6 1/4
6. 2 1/2
7. 8 1/2
8. 1 7/9
9. 6 2/5
10. 1 2/3
Problem Solving: $42

PAGE 14
Review Exercises:
1. 1 4/5
2. 4/5
3. 2 2/9
4. 21
5. 2/8, 1/4
6. 8
S1. 5/2
S2. 19/3
1. 36/5
2. 11/3
3. 21/4
4. 19/4
5. 17/6
6. 57/8
7. 23/3
8. 35/4
9. 37/4
10. 15/8
Problem Solving: 18 gallons

PAGE 15
Review Exercises:
1. 33
2. 9
3. 2 1/7
4. 4/5
5. answers vary
6. 2/3
S1. 17/2
S2. 19/4
1. 15/2
2. 17/3
3. 43/8
4. 8/3
5. 28/3
6. 38/5
7. 23/3
8. 33/2
9. 23/5
10. 23/4
Problem Solving: 6 packages

Solutions

PAGE 16
Review Exercises:
1. 1 3/7
2. 23/5
3. 5/7
4. 21/4
5. 1 1/3
6. answers vary
S1. 4/5
S2. 1 1/4
1. 7/9
2. 3/4
3. 1 1/2
4. 9/10
5. 1 1/4
6. 1 1/4
7. 1 4/5
8. 1 1/4
9. 1 3/5
10. 1 2/3
Problem Solving: 1 1/4 cups

PAGE 17
Review Exercises:
1. 11/2
2. 3 1/5
3. 4/5
4. 1 1/2
5. answers vary
6. 9/10
S1. 4/5
S2. 1 1/2
1. 3/4
2. 7/10
3. 1 2/5
4. 1 1/8
5. 1 1/12
6. 1 3/4
7. 1 2/3
8. 1 3/13
9. 2
10. 1 1/5
Problem Solving: 1 3/5 inches

PAGE 18
Review Exercises:
1. 1 1/8
2. 3/5
3. 8/9
4. 5 1/4
5. 15/4
6. 1
S1. 6 4/5
S2. 8 1/4
1. 7 6/7
2. 6 1/5
3. 12
4. 6 1/3
5. 7 2/7
6. 7 1/2
7. 10 1/5
8. 6 2/15
9. 10
10. 12 1/5
Problem Solving: 1 3/4 cups

PAGE 19
Review Exercises:
1. 2/3
2. 4/5
3. 1 1/5
4. 15/16
5. 23/6
6. 5 2/3
S1. 7 4/5
S2. 9 3/5
1. 5 4/5
2. 8 2/5
3. 15
4. 13 1/2
5. 30 1/5
6. 17 1/3
7. 8 1/15
8. 5 1/2
9. 11
10. 13 5/11
Problem Solving: 1/2 mile

PAGE 20
Review Exercises:
1. 1 1/2
2. 4/5
3. 1 2/5
4. 1 1/4
5. 4/5
6. 9 4/7
S1. 1 1/4
S2. 8 1/5
1. 2/3
2. 8
3. 1 3/5
4. 9 1/3
5. 11 1/2
6. 1 2/3
7. 5/8
8. 13
9. 1 1/4
10. 6 1/3
Problem Solving: 14 1/2 hours

PAGE 21
Review Exercises:
1. answers vary
2. 3 1/2
3. 7/10
4. 1 1/3
5. 1
6. 16 1/2
S1. 7 6/7
S2. 1 1/2
1. 1/2
2. 9 1/4
3. 1 7/8
4. 13 1/3
5. 4/5
6. 10 1/5
7. 1 1/6
8. 13 2/3
9. 11 7/8
10. 1 1/9
Problem Solving: 90

PAGE 22
Review Exercises:
1. 1/2
2. 1 2/7
3. 1 1/4
4. 8
5. 15 4/5
6. 10 1/2
S1. 1/2
S2. 5/8
1. 3/5
2. 3/5
3. 3/4
4. 3/4
5. 7/9
6. 1/2
7. 4/5
8. 2/7
9. 3/4
10. 1/5
Problem Solving: 1/2 pounds

PAGE 23
Review Exercises:
1. 4/5
2. 9 2/5
3. 23/3
4. 4/5
5. 1 2/5
6. 2 5/7
S1. 1/2
S2. 7/8
1. 3/4
2. 4/5
3. 1/4
4. 1/3
5. 3/4
6. 1/3
7. 1/3
8. 3/5
9. 4/5
10. 1/4
Problem Solving: 10 3/4 hours

PAGE 24
Review Exercises:
1. 1/2
2. 10
3. 1/4
4. 8
5. 26 1/3
6. 1 5/8
S1. 3 1/2
S2. 2 3/5
1. 4 1/4
2. 4 1/2
3. 3 5/8
4. 5 3/4
5. 2 3/5
6. 2 3/5
7. 3
8. 1 2/3
9. 4 5/7
10. 5 3/5
Problem Solving: 11 1/2 miles

PAGE 25
Review Exercises:
1. 1 4/15
2. 3/4
3. 15/16
4. 1 1/7
5. 3/15, 1/5
6. 8 1/4
S1. 1 1/2
S2. 5 1/2
1. 6 1/3
2. 5 2/3
3. 4 3/4
4. 3 1/2
5. 2 8/11
6. 4 5/6
7. 6 1/2
8. 4 3/4
9. 4 7/10
10. 3 1/3
Problem Solving: 4 Gallons, $12

PAGE 26
Review Exercises:
1. 3/4
2. 2 3/4
3. 3 3/4
4. 1 1/2
5. 6 4/5
6. 9 2/15
S1. 3 2/5
S2. 8 3/8
1. 4 7/8
2. 2 4/7
3. 11 8/15
4. 4 1/10
5. 12 7/8
6. 12 6/7
7. 11 2/11
8. 5 4/5
9. 23 4/7
10. 15 8/15
Problem Solving: 9 1/4 yards

PAGE 27
Review Exercises:
1. answers vary
2. 4 2/3
3. 5 3/4
4. 1 3/4
5. 3/5
6. 19/5
S1. 8 1/8
S2. 9 4/7
1. 4 3/4
2. 32 5/8
3. 3 1/16
4. 8 3/8
5. 10 23/24
6. 15 2/5
7. 39 11/24
8. 8 11/24
9. 12 1/20
10. 24 1/5
Problem Solving: 11 1/2 gallons

Solutions

PAGE 28
Review Exercises:
1. 4 2/3
2. 9 2/15
3. 1 6/25
4. 3/5
5. 4 1/4
6. 1 1/6
S1. 6 1/5
S2. 3 3/5
1. 1 2/7
2. 2/3
3. 11 2/9
4. 6 3/4
5. 9 3/4
6. 6 1/2
7. 8 1/3
8. 4 2/3
9. 7 3/5
10. 1 9/10
Problem Solving: 450 miles per hour

PAGE 29
Review Exercises:
1. 2 1/2
2. 6 4/15
3. 17 2/5
4. 1 1/2
5. 5/8
6. 21/4
S1. 4 3/4
S2. 8 1/2
1. 2/3
2. 1 4/9
3. 11 1/7
4. 6 2/5
5. 68 1/4
6. 3 1/4
7. 14 1/5
8. 4 1/2
9. 11
10. 2
Problem Solving: 79 desks

PAGE 30
Review Exercises:
1. 1 1/4
2. 1 1/2
3. 7/8
4. 17/5
5. 1 2/3
6. 4 1/4
S1. 20
S2. 24
1. 18
2. 12
3. 21
4. 24
5. 36
6. 56
7. 48
8. 60
9. 48
10. 48
Problem Solving: 26 students

PAGE 31
Review Exercises:
1. 1 1/5
2. 16 1/4
3. 9 7/8
4. 2/5
5. 6 3/8
6. 5 1/3
S1. 24
S2. 12
1. 18
2. 28
3. 30
4. 90
5. 36
6. 36
7. 12
8. 48
9. 72
10. 30
Problem Solving: 3 1/4 gallons

PAGE 32
Review Exercises:
1. 1 1/3
2. 1/4
3. 30
4. 4/5
5. 5 2/11
6. 7 1/2
S1. 7/12
S2. 1 3/10
1. 8/9
2. 1 1/6
3. 11/12
4. 1 5/12
5. 1 1/4
6. 1 1/4
7. 11/12
8. 1 1/36
9. 1 3/22
10. 13/24
Problem Solving: 1 1/2 hours

PAGE 33
Review Exercises:
1. 1 1/5
2. 8 1/4
3. 4 3/10
4. 2 5/8
5. 4 1/2
6. 24
S1. 7/9
S2. 1 7/12
1. 11/12
2. 1 3/20
3. 22/25
4. 1 5/12
5. 1 1/15
6. 1 4/15
7. 1 13/24
8. 14/33
9. 19/24
10. 23/30
Problem Solving: $36

Solutions

PAGE 34
Review Exercises:
1. 3/5
2. 1
3. 10 2/5
4. 11/12
5. 1 3/20
6. 1 1/4
S1. 2/9
S2. 7/12
1. 3/40
2. 17/30
3. 19/30
4. 1/4
5. 1/6
6. 7/18
7. 1/4
8. 3/10
9. 33/56
10. 2/5
Problem Solving: 4 3/4 pounds

PAGE 35
Review Exercises:
1. 3/5
2. 3/4
3. 1 2/3
4. 4 1/12
5. 8 1/4
6. 19/30
S1. 1/4
S2. 27/40
1. 1/2
2. 9/16
3. 3/8
4. 3/4
5. 5/12
6. 13/18
7. 3/10
8. 2/5
9. 1/12
10. 17/24
Problem Solving: 7 boxes,
9 left over

PAGE 36
Review Exercises:
1. 1 3/4
2. 1 1/15
3. 13/15
4. 1/4
5. 4 2/5
6. 2/15
S1. 5/8
S2. 1 3/8
1. 8/9
2. 1/12
3. 3/5
4. 1 3/10
5. 13/22
6. 1/2
7. 1/2
8. 21/25
9. 5/14
10. 17/20
Problem Solving: 1 5/12 pounds

PAGE 37
Review Exercises:
1. 8 1/2
2. 12 2/5
3. 10 3/5
4. 4 1/2
5. 3 2/5
6. 8 1/5
S1. 1/2
S2. 1 1/12
1. 3/4
2. 3/4
3. 13/30
4. 1/16
5. 17/24
6. 11/15
7. 1/8
8. 7/8
9. 7/10
10. 1 2/9
Problem Solving: 4 1/4 hours

PAGE 38
Review Exercises:
1. 23/7
2. 1 2/15
3. 4/7
4. answers vary
5. 1 1/8
6. 1 1/8
S1. 7 11/15
S2. 8 1/12
1. 5 1/4
2. 7 5/6
3. 7 9/10
4. 8 1/5
5. 11 2/9
6. 7 1/10
7. 8 3/14
8. 11 1/18
9. 7 10/21
10. 11 5/24
Problem Solving: 17 1/4 inches

PAGE 39
Review Exercises:
1. 1 1/12
2. 1 1/21
3. 1 5/16
4. 3/10
5. 1/2
6. 1/2
S1. 9 1/4
S2. 8 1/6
1. 11 1/4
2. 8 9/10
3. 10 1/6
4. 12 13/14
5. 5 1/8
6. 16 1/6
7. 7 3/20
8. 14 1/10
9. 5 7/9
10. 8 1/32
Problem Solving: 9 1/4 dollars

PAGE 40
Review Exercises:
1. 3/4
2. 4 9/16
3. 3 1/2
4. 3 2/3
5. 2 1/3
6. 1/2
S1. 3 1/6
S2. 3 5/6
1. 5 1/4
2. 6 7/10
3. 3 8/15
4. 1 7/10
5. 3 11/14
6. 5 3/10
7. 2 4/9
8. 3 9/16
9. 1 11/12
10. 2 11/15
Problem Solving: 16 1/4 dollars

PAGE 41
Review Exercises:
1. 5 3/5
2. 13 1/2
3. 9 3/4
4. 11 14/15
5. 11 1/16
6. 1 1/30
S1. 3 3/4
S2. 2 4/5
1. 4 1/6
2. 4 5/6
3. 2 8/9
4. 2 5/6
5. 5 7/16
6. 3 11/15
7. 8 3/10
8. 3 1/8
9. 4 9/10
10. 6 11/24
Problem Solving: 22 1/4 dollars

PAGE 42
Review Exercises:
1. 5/6
2. 5/8
3. 2 3/4
4. 4
5. 6 5/6
6. 11 1/8
S1. 4/5
S2. 2 1/2
1. 13/24
2. 13/18
3. 3 4/7
4. 13 7/8
5. 18 3/5
6. 4 7/9
7. 3 7/20
8. 3 1/2
9. 23/30
10. 10 1/2
Problem Solving: $204

PAGE 43
Review Exercises:
1. 3/5, 6/10
2. 15/2
3. 21
4. 6 1/5
5. 4/5
6. 1 4/5
S1. 2 1/2
S2. 5 1/4
1. 7/8
2. 1 1/4
3. 17/24
4. 2 1/2
5. 12
6. 6 7/12
7. 12 3/14
8. 1 7/60
9. 1 1/8
10. 8 3/10
Problem Solving: 33 3/4 minutes

PAGE 44
Review Exercises:
1. 5 5/6
2. 3 5/6
3. 1 3/10
4. 4 1/16
5. 7 5/6
6. 2/3
S1. 15/28
S2. 5/9
1. 5/63
2. 1/5
3. 1 7/8
4. 10/27
5. 1 1/15
6. 1 1/15
7. 2 1/4
8. 32/35
9. 2/5
10. 1 13/14
Problem Solving: 37 1/2 pounds

PAGE 45
Review Exercises:
1. 19/8
2. 1/2
3. 4 1/2
4. 13/15
5. 1 2/5
6. 1 4/5
S1. 3/5
S2. 2 5/8
1. 8/15
2. 1 1/4
3. 1 7/20
4. 1 1/3
5. 5/8
6. 1 7/8
7. 3 1/2
8. 5/9
9. 1 1/4
10. 2/3
Problem Solving: $17

PAGE 46
Review Exercises:
1. 3/8
2. 1 2/3
3. 1 1/10
4. 3 1/6
5. 3 3/4
6. 6 2/3
S1. 7/15
S2. 2/3
1. 9/16
2. 1/12
3. 11/18
4. 3 2/3
5. 3/10
6. 2/3
7. 2/5
8. 9/20
9. 1 3/7
10. 1 1/21
Problem Solving: 83 seats

PAGE 47
Review Exercises:
1. 1 2/3
2. 5/8
3. 1 1/6
4. 25/2
5. 6 1/6
6. 7/16
S1. 2/5
S2. 1
1. 7/20
2. 20/27
3. 11/24
4. 4
5. 2 1/4
6. 1/4
7. 1 1/2
8. 1 1/4
9. 1 1/3
10. 11/15
Problem Solving: $13,000

PAGE 48
Review Exercises:
1. 5/6
2. 13 1/2
3. 29/5
4. 12/17
5. 7/20
6. 2 1/6
S1. 12
S2. 3 3/4
1. 12
2. 10
3. 18
4. 2 1/7
5. 18 1/2
6. 1 1/2
7. 2 1/3
8. 7 1/2
9. 35
10. 4 1/5
Problem Solving: 14 girls

PAGE 49
Review Exercises:
1. 1 3/5
2. 8 1/3
3. 10 5/12
4. 2/3
5. 1/4
6. 3/8
S1. 5 3/5
S2. 25
1. 20
2. 9 3/5
3. 6 2/3
4. 45
5. 3 3/4
6. 20
7. 35
8. 16 2/3
9. 4 4/5
10. 3 1/3
Problem Solving: 180 girls

PAGE 50
Review Exercises:
1. 10/21
2. 10/13
3. 6
4. 10
5. 4 2/3
6. 24
S1. 4/9
S2. 5 1/4
1. 1 1/6
2. 7 1/3
3. 11
4. 4 2/5
5. 8 1/4
6. 8 1/8
7. 8 3/4
8. 20
9. 16 1/3
10. 2 5/7
Problem Solving: 36 miles

PAGE 51
Review Exercises:
1. 7/18
2. 3 3/4
3. 16
4. 6 1/6
5. 2 14/15
6. 8 1/10
S1. 4
S2. 6
1. 2 1/4
2. 6 7/8
3. 8 3/4
4. 24
5. 14 2/3
6. 3 3/8
7. 11 1/4
8. 8 2/3
9. 8 1/6
10. 10
Problem Solving: 275 cars

Solutions

PAGE 52
Review Exercises:
1. 7/12
2. 12
3. 1 2/3
4. 17 1/2
5. 3
6. 3 2/3
S1. 7/18
S2. 4
1. 3/10
2. 1 7/11
3. 27
4. 1 7/8
5. 2 1/10
6. 8/9
7. 11
8. 2 2/15
9. 20
10. 12 3/5
Problem Solving: $60

PAGE 53
Review Exercises:
1. 1/4
2. 13/24
3. 4 1/15
4. 4 2/5
5. 8 7/12
6. 1 7/10
S1. 2 1/2
S2. 7 7/8
1. 1/4
2. 2 1/4
3. 9
4. 9/10
5. 6 3/4
6. 150
7. 5 1/3
8. 22 1/2
9. 5 1/4
10. 21
Problem Solving: 24 students

PAGE 54
Review Exercises:
1. 9/10
2. 5/18
3. 9 1/8
4. 12 1/2
5. 1 3/5
6. 4 1/2
S1. 1 1/2
S2. 2/7
1. 1/7
2. 1 3/5
3. 4/17
4. 1/15
5. 3 1/2
6. 9
7. 1/12
8. 4 1/2
9. 2/11
10. 1/17
Problem Solving: $75

PAGE 55
Review Exercises:
1. 15/2
2. 9/20
3. 2 2/7
4. 49
5. answers vary
6. 1/3, 3/9
S1. 1 1/7
S2. 4/19
1. 1 4/11
2. 1/6
3. 9/28
4. 1/16
5. 7 1/2
6. 6
7. 1/50
8. 4
9. 3/23
10. 1 5/11
Problem Solving: 2 1/4 degrees

PAGE 56
Review Exercises:
1. 1 5/12
2. 1 1/2
3. 4 2/3
4. 5 1/2
5. 1 5/12
6. 1 1/4
S1. 1 3/5
S2. 2 1/4
1. 3 3/4
2. 3/4
3. 10 1/2
4. 11
5. 4 2/3
6. 3 6/11
7. 4/9
8. 3 3/4
9. 3
10. 1 13/15
Problem Solving: 9 pieces

PAGE 57
Review Exercises:
1. 7/12
2. 3
3. 5
4. 1 1/4
5. 1
6. 1 7/9
S1. 1 1/6
S2. 2
1. 2 1/2
2. 6/7
3. 3 1/3
4. 26
5. 3 2/3
6. 3 3/8
7. 8
8. 1 7/8
9. 1 11/16
10. 5
Problem Solving: 27 inches

Solutions

PAGE 58
Review Exercises:
1. 1 1/8
2. 3/16
3. 1 1/8
4. 3 1/2
5. 4 2/3
6. 2
S1. 6/7
S2. 5/6
1. 11/12
2. 3/5
3. 5/6
4. 9/11
5. 13/15
6. 3/8
7. 23/30
8. 2/11
9. 9/10
10. 11/12
Problem Solving: 19 miles

PAGE 59
Review Exercises:
1. 1 1/2
2. 10
3. 3
4. 9/10
5. 6 7/8
6. 10 1/2
S1. 7/9
S2. 11/20
1. 11/14
2. 10/12
3. 3/4
4. 1/2
5. 2/6
6. 11/20
7. 5/6
8. 7/8
9. 3/5
10. 3/8
Problem Solving: 23

PAGE 60
1. 4/5
2. 4/5
3. 5/6
4. 2 1/5
5. 2 1/3
6. 3 12/25
7. 28/5
8. 64/5
9. 22/3
10. 40
11. 42
12. 60
13. 12
14. 60
15. 12
16. 9/15, 3/5
17. 6/8, 3/4
18. answers vary
19. 2/3
20. 5/6

PAGE 61
1. 1
2. 1 1/2
3. 1 3/20
4. 6 3/8
5. 9 1/2
6. 3/4
7. 2 5/7
8. 4 3/7
9. 4 1/4
10. 4 8/15
11. 3/14
12. 7/32
13. 24
14. 3 1/2
15. 8 3/4
16. 2 1/2
17. 14
18. 1 1/2
19. 2 8/11
20. 2 3/11

PAGE 62
1. 6/7
2. 3/5
3. 4/5
4. 3 1/4
5. 1 7/12
6. 1 3/7
7. 17/8
8. 26/3
9. 17/16
10. 9
11. 16
12. 49
13. 30
14. 24
15. 30
16. 2/6, 1/3
17. 2/8, 1/4
18. answers vary
19. 4/7
20. 4/5

PAGE 63
1. 4/5
2. 1 1/2
3. 1 3/10
4. 6 1/15
5. 8 2/21
6. 4/5
7. 3 3/4
8. 3 5/8
9. 4 11/20
10. 4 2/3
11. 4/9
12. 1/6
13. 21
14. 2 3/5
15. 6
16. 1 1/2
17. 7 1/3
18. 1 13/14
19. 2 6/17
20. 1 1/2

Math Notes

Math Notes

Math Notes

91762986R00046

Made in the USA
Middletown, DE
02 October 2018